Contested Waters

Contested Waters

Sub-national Scale Water and Conflict in Pakistan

Daanish Mustafa

BLOOMSBURY ACADEMIC
LONDON · NEW YORK · OXFORD · NEW DELHI · SYDNEY

BLOOMSBURY ACADEMIC
Bloomsbury Publishing Plc
50 Bedford Square, London, WC1B 3DP, UK
1385 Broadway, New York, NY 10018, USA
29 Earlsfort Terrace, Dublin 2, Ireland

BLOOMSBURY, BLOOMSBURY ACADEMIC and the Diana logo
are trademarks of Bloomsbury Publishing Plc

First published in Great Britain 2021
This paperback edition published in 2022

Copyright © Daanish Mustafa, 2021

Daanish Mustafa has asserted his right under the Copyright, Designs and Patents Act, 1988, to be identified as Author of this work.

For legal purposes the Acknowledgements on p. xi constitute an extension of this copyright page.

Series design by Adriana Brioso

All rights reserved. No part of this publication may be reproduced or transmitted in any form or by any means, electronic or mechanical, including photocopying, recording, or any information storage or retrieval system, without prior permission in writing from the publishers.

Bloomsbury Publishing Plc does not have any control over, or responsibility for, any third-party websites referred to or in this book. All internet addresses given in this book were correct at the time of going to press. The author and publisher regret any inconvenience caused if addresses have changed or sites have ceased to exist, but can accept no responsibility for any such changes.

A catalogue record for this book is available from the British Library.

A catalog record for this book is available from the Library of Congress.

ISBN: HB: 978-1-7883-1342-1
PB: 978-0-7556-3523-8
ePDF: 978-0-7556-3521-4
eBook: 978-0-7556-3520-7

Typeset by Deanta Global Publishing Services, Chennai, India

To find out more about our authors and books visit www.bloomsbury.com and sign up for our newsletters

To Kashmir

Contents

List of illustrations	ix
Acknowledgements	xi

1 Introduction: Contested waters in hydro-hazardscapes   1
   What is in the scale?   4
   Of expert visions, power and conflict   6
   Water conflict, in a hazardous world   8
   Pakistan as the most water scarce/vulnerable country,
      and what does it matter?   12
   How is this book organized   18

2 Nationalist hazardscapes: The case of inter-provincial water conflict   21
   Historical antecedents of inter-provincial water conflict   23
   Politico-historic context of inter-provincial relations in Pakistan   25
   Agreements, allocations and royalties in inter-provincial
      water politics   28
   Kalabagh or death/death or Kalabagh: Competing narratives
      of inter-provincial water conflict in Pakistan   34
   Conclusion: Mitigating inter-provincial water conflict   40

3 Local-scale water conflict over surface and groundwater in
   rural Pakistan   43
   The historico-institutional context of canal and Karez
      irrigation in Pakistan   44
   Pulling the local into global: Groundwater conflict in Balochistan   47
   Producing the local from national: Surface irrigation water conflict   50
   Conclusion and future pathways   53

4 Contested hazards in local hazardscapes: From floods to pollution   57
   The historico-physical geography of floods in Pakistan   58
   Drainage as a hazard   61
   The fractal scalar politics in flood and drainage management
      in Pakistan   63

	Contrasting flood conflicts in Sindh and Punjab	67
	Downstream conflicts from pollution hazard	71
	Conclusion	77
5	Conflict over domestic water supply: The case of Karachi	81
	Physical and institutional context of water supply in Karachi	82
	Privatization by other means: How it never did and can work for Karachi's poor	83
	Have power? Have water	88
	Conclusion: Possible ways forward	96
6	Conclusion	99
	References	105
	Index	110

Illustrations

Figures

1.1	The Indus River Basin and its major infrastructure	13
1.2	Comparative silt loads of major rivers of the world	14
2.1	Flows in MAF downstream of Kotri (PILDAT, 2011)	31
2.2	A poster along the Mall Road, Lahore, urging Pakistan's popular chief of the army staff (COAS), General Raheel Sharif to undertake the construction of Kalabagh Dam	35
3.1	Schematic diagram of a Karez	45
4.1	The bed of the Ravi River downstream of the Sidhnai Barrage with extensive silt deposition within the channel	59
4.2	A map of the inundation zone on the right bank of the Ravi River	60
4.3	The 2010 inundation situation in Muzaffargarh district clearly showing the road acting as a barrier to the drainage of the flood peak	61
4.4	Location of Manchar Lake in the Sindh province of Pakistan	63
4.5	A view of the Khan Mohammad Mallah Village, District Sewan	72
4.6	Focus Group Discussion at Khan Mohammad Mallah village	73
4.7	Women scooping up brackish water from a groundwater source	74
5.1	Overhead pipe at the Altaf Nagar pumping station, Karachi	85
5.2	Tanker trucks lined up to fill up at the Germany pump. This is the closest we could get to the pumping station safely	86
5.3	Little boys carrying water bottles from a water vendor in Orangi	87
5.4	Corner meeting with residents of Baldia Town	91
5.5	Community meeting in Bhagwandas Goth	92
5.6	Sewerage line manholes in Bhagwandas	93
5.7	Water delivered by a tanker in Bhagwandas. This water is used for drinking as well	94
5.8	Girls knocking on the door to request water	95

All photographs have been taken by the author and all other figures have either been made/commissioned by the author.

Tables

2.1 Agreed Apportionment of Water according to the Inter-provincial Water Accord 1991 (figures in MAF)   25
4.1 Control Station Design Discharge Capacities   67

Acknowledgements

The research presented in this book was funded by the United States Institute for Peace (USIP). I thank Moeed Yusuf for initiating and supporting the idea for the research and Colin Cookman at USIP for his useful feedback as the research progressed. A great debt of gratitude is owed to Giovanna Gioli for her energy, intellect and leadership in the conduct of the research for this book. Milan Karner and Imran Khan were integral parts of the research team, and I am grateful for their contributions. I gratefully acknowledge Zarif Khero, Shamsuddin, Syed Faisal Imam (Late), Simi Kamal, Sono Khangrani, Manzoor Memon, Muhammad Asif Iqbal and Zuhair Ashir in addition to the staff members of the Hisaar Foundation and Social Policy and Development Centre (SPDC), for facilitating the fieldwork for the research presented here. David Stonestreet was always an amazing colleague and I thank him for commissioning this manuscript, as I acknowledge Tomasz Hoskins's and Nayiri Kendir's help during the publication process. Lastly, I affectionately thank my parents Rashid and Farhat, along with Zaigham, Erum, Shahzaman and Mustafa in addition to too many friends, colleagues and well-wishers to name, for their continued faith in me, and my work. I alone am responsible for any omissions or errors in this book.

1

Introduction

Contested waters in hydro-hazardscapes

Water is creative. Water is productive. But what is water? Asks Linton (2010). In a lab it can be H$_2$O, but that's hardly relevant to the universe of material functionalities and discursive meanings that societies ascribe to this compound. Within seconds of its escape from the rarefied confines of a chemistry lab, water bonds with a host of other molecules and takes on the materiality that makes it the bases of ecology, society, politics and life itself. The problematique for us is not so much what is water, but rather how we produce ourselves through water. And produce ourselves we do, as men, women, transgender, farmers, fisherfolk, artisans, majorities, minorities, patriots, pioneers, ethno-nationalists, indigenous, modern and traditionalists. In this account of conflict over water at, and across, geographical scales, we demonstrate how conflicts over water are not just over water. In fact, the conflicts are also often through water, in its many forms about meaning and visions of social life. Drawing upon case studies of water conflict at the sub-national scale in Pakistan, we seek to establish how water is productive, not only of identities, developmental visions, modernity and society but of geographical scale itself.

Transboundary water conflict is perhaps the most well-known and researched form of water conflict in the world, particularly since the turn of the twentieth century. In the twentieth century the Westphalian nation state system, with its territorial sovereignty and defined borders, really came into its own. One of the earliest modern transboundary water treaties was between the United States and Mexico on the Colorado River, which defined a specific volumetric right of the lower riparian Mexico, in 1944. The 1944 Treaty too had the task of balancing the water allocations to the US states, with the US government's sovereign commitment to delivering a fixed quantity of water, downstream to Mexico. The water to be delivered to Mexico, it turned out, did not exist in the river because of over allocation to the riparian states through the Colorado River

Compact 1922. Here, from the very inception of the transboundary water conflict and negotiations, the sub-national actors, namely the six riparian states of the Colorado – chiefly California – had a significant influence on how the treaty was negotiated and how it was to be administered over subsequent years (Bates et al., 1993; Moore, 2018). We centre this tension between the sub-national and the international in our narrative.

As the post-colonial world started to take shape with its arbitrarily drawn national boundaries and jealous nationalisms, the transboundary water issue too started becoming prominent. The signature water conflicts in the middle of the twentieth century were on the Indus River between India and Pakistan, and on the Jordan, Nile and Tigris and Euphrates Rivers in Middle East and North Africa (MENA) region. In case of the Indus, a resolution of sorts was facilitated by the World Bank and Western powers in the form of the Indus Waters Treaty (IWT), dividing the basin's water between the two countries. Pakistan, however, was able to negotiate and agree to the IWT, which was not exactly to its advantage (Akhter 2015 Alam 2002), partially because of the promise of massive infrastructure investment, and partially because it had silenced the potentially oppositional voices by dissolving its federating units into a single (One Unit) entity of West Pakistan (Michel, 1967). This constitutional sleight of hand handed Pakistan the IWT, but down the road created a number of issues of inter-provincial water conflict for the federation of Pakistan, as we will elucidate in Chapter 2 of this book.

The water conflicts in the MENA region spawned some of the most sophisticated literature on transboundary water conflict, which in the first instance soundly broke away from the 'Water Wars' thesis that had come to dominate the international discourse on water in the post–cold war era. The Water Wars thesis being the simplistic notion in the 1990s that where countries fought over ideology and resources in the past, the future armed conflicts will be over water (e.g. see Homer-Dixon, 1999). While the thesis gained considerable traction in the policy and popular discourse, it was quickly debunked by systematic and statistical research which demonstrated that countries are more likely to cooperate over water than fight (e.g. see Wolf, 2002; Brochmann & Gleditsch, 2012). Zeitoun and Warner (2006) went further and proposed the hydro-hegemony framework for understanding how apparent cooperation doesn't always lead to fair outcomes as a stronger riparian may negotiate favourable water agreements to itself by virtue of its superior military and, more importantly, power/knowledge position. Recognizing this simultaneous existence of conflict and cooperation, Zeitoun and Mirumachi (2008) developed a Transboundary Waters Interaction Nexus (TWINS) to systematically evaluate the balance between different levels of conflict and cooperation in different contexts. Moore (2018), in conversation with the transboundary water conflict literature, and consistent with this book, argues that sub-national scale water conflict is likely to be a much more prevalent form of hydropolitics than transboundary conflict. Moore, nevertheless, engages with water conflict as necessarily a negative phenomenon, whereas in this book I maintain that conflict is not necessarily a bad thing. In fact, at times it is necessary in the cause of justice, and definitely more preferable to unjust, hegemonic cooperation.

The transboundary discourse, however, is predicated upon a fundamental scientific abstraction of cubic metres or acre feet of water, typically expressed in millions of cubic metres (MCM) or acre feet (MAF). Critical as these units of measurement may be to the national- and international-scale water policy and politics, they do also transmute water into a national- and international-scale material to be contested at that scale. I argue through the course of this book that MCMs and MAFs, therefore, also become the building blocks for the construction of national and international scales. My concern is to destabilize the scalar hegemony of cubic metres and acre feet and instead link water to its political context at and across human scales of experience, emotions, memories and livelihoods. This is not an unprecedented ambition. Moore (2018) too suggests sectional identities, decentralization and political opportunity structures as the key drivers of sub-national scale hydropolitics and the conflict inherent therein. While I share Moore's concern with the imbrication of hydropolitics with sectional identities, I move beyond his jurisdictional approach to sub-national hydropolitics where sub-national provinces or states act like international nation states. I trace the dynamics of sub-national water conflict across local/municipal/watercourse scale to regional and provincial scale. I, furthermore, also consider conflict, not just over water's quantitative distribution but also on the distribution of its hazards.

I have often, with considerable trepidation, answered to the honorific of a water expert. Expertise is typically synonymous with technocracy, with scientific or managerial mandarins abstracting the materiality of things like water into high-minded concepts and measuring units. I want to distance myself from that technocratic register of engaging with water, and instead aim to tell the tale of water and conflict in a distinctly non-technocratic and hence non-expert register. But non-technocratic also implies expressly political and hence theoretically grounded mode. Given the diffuse and multifaceted materiality of water, it is one medium that defies expertise; hence I don't claim any. I hope that as I try to frame and tell the story of human interactions with water, I speak to the experiential and philosophical interactions of all the readers of this book regardless of their disciplinary training or background.

What is in the scale?

The concept of geographical scale has been integral to popular conversations and academic research on water. From basin, to transboundary, to interprovincial, to local, the concept of scale lies at the heart of the water conflict problematique (Marston et al., 2005; Hoefle, 2006). Moore (2018) too considers scale to be the central dilemma of sub-national scale hydropolitics, but he doesn't engage with any of the geographical conceptions of space, and instead proposes functional, vertical and horizontal jurisdictional tensions between water and human institutions to define interjurisdictional cooperation as the fundamental dilemma of water conflict at the sub-national scale (p. 38). I hope to move beyond the simple institutional jurisdictions' lens towards scale, to ask a more open question: What is geographical scale? For some it is simply the vertical structure indicating the size of the spatial territorial unit being referred to, for example, local, national, global – obvious, material and measurable. In its 'obvious' physical sense the scalar unit becomes the spatial container and context for social relations to actuate (Hoefle, 2006). On the other hand, and in a more critical sense, scale is not an ontological given, but rather a useful/intuitive epistemological tool to apprehend spatial interactions, albeit in a vertical structure. In this later conceptualization it is, well, just a conceptual tool, historically contingent and hence opening up greater possibilities for apprehending how space and its accompanying scales are in fact produced politically (Swyngedouw, 2004; Leitner & Miller, 2007).

Marston et al. (2005) – not quite satisfied with the notion that scale is in fact produced through politics, or as politics produce space they also produce scale – take exception to inevitable verticality implied in the concept of scale. Drawing primarily upon Deleuze and Gottari, they critique the inherently vertical visions of the world implied in the concept of scale and instead argue for dispensing with the concept altogether. They argue that instead of vertical spatial imaginaries that eviscerate agency in the name of faceless global scale structures, one needs a more horizontal dispersed imaginary that celebrates the emergent, the fluid, the local and the differentiated relations therein.

I do see some merit in Marston et al.'s (2005) critique of vertical imaginaries and of structural thinking that inheres in the concept of scale. I, nevertheless, am not entirely comfortable with their call for dispensing with the concept of scale altogether, or the irrelevance of even recognizing some vertical hierarchies in the social world. Instead, I find myself in greater agreement with Hoefle (2006), Jonas (2006) and Leitner and Miller (2007), who argue that, of course, scale is not an ontological given. In fact, ontology and epistemology are coproduced, and to have one without the other is problematic at best. Second, discussions of scale are really discussions about power, and how power relations spawn spaces and spatial hierarchies to reproduce power structures. To turn away from the question of power is to renege on the fundamental covenant of political and politicized inquiry. And lastly, engagement with the concept of scale does not preclude or devalue agency. If anything, it adds nuance and relief to how emancipatory alliances form across space, to contest hegemonic spatial practices and representations.

To the above defence of the relevance of scale, one must also add that verticality does not imply a one-way flow of power or causality. Much of the evidence I present in this book is also about how agency and power can also flow in multiple directions. I view scale as a socially constructed spatial structure, which is helpful in understanding spatial practices and representations of power. I do not view scale as a container of social relations but rather as an outcome of the same. In the same vein, then, I seek to understand how conflict over water access and distribution of its hazards may actuate, be negotiated and distributed across space.

In the project of understanding conflict over water access and hazards, I find Qutub et al.'s (2003) insight on the fractal nature of conflict over water across spatial scales to be quite insightful. Drawing upon case studies across local, intrastate and transboundary water conflicts in South Asia, they discern a fractal pattern of repetition of issues, themes, actors, interests and strategies within and across scalar boundaries. Although they do not quite theorize the fractal nature of scale and conflict in their studies, we find the insight eminently useful in enabling our thinking on scale and its relationship to water conflict.

The term 'fractal' is perhaps one of the most vexingly difficult concepts to define in mathematics and even in common usage. Instead of rehearsing the mathematical debates surrounding the nature of fractals, suffice it to say that for the present purposes one may think of 'fractal' as denoting the notion of self-similar complex patterns repeating themselves through space characterized by varied detail across scales – spatial scales for our purposes. The notion of self-similarity to my mind hearkens to the concern with social power as the common denominator in any water conflict at or across spatial scales, inducing the self-similar patterns of conflict at all scales. The fractal variance and complexity of water conflict across scales is consonant with the uncertainty borne of water's own fluid materiality and the complexity of the social system within which it is embedded. With the above insight in mind we turn to the questions of power and then uncertainty, which frame the theoretical framework for our investigation into the fractal nature of water conflict. But first to power.

Of expert visions, power and conflict

The question of social power has been a presence at the heart of social scientific inquiry for more than two centuries. Sometimes it has been engaged with in a political economic register and at other times as per Foucault (1980) in a power/knowledge mode, where power permeates human bodies and polities at the capillary level. My own earlier entanglements with the concept of power were in a critical Realist mode, where drawing upon Chatterjee (1983) I had defined power as structurally emanating from the feudal, bourgeoise and communal modes of social organization. The empirical manifestations of each of these structural modes of power were violence, ability to compensate and hegemonic power/knowledge. My argument at the time was that the degree of access to irrigation water and vulnerability to flood hazard was contingent upon the operation of these different modes of power. The core argument being that it was not physical geography but rather power geographies that determined who was more vulnerable, or who got more or less of the irrigation water (Mustafa, 2002).

Within hazards research, Hewitt (1983) earlier on noted that hazards and risks themselves were produced through technocratic modes of governance and thinking. The very notion of hazards and disasters as unplanned events removed from the chronic hazardousness of everyday life, he argued, was a dangerous fiction, which detracted from the injustices and structural problems that made people susceptible to suffer damage from environmental extremes (Hewitt, 1983). More specifically to the context of the Indus, Gilmartin (1994) described the tension between 'imperial science and science of the empire' – the two representing the notions of scientific control over nature and the 'scientific discourse of power', respectively (Gilmartin, 1994, p. 1127). The scientific discourse of power in this instance was based upon imperial racialized categories and political calculations on how to maintain imperial control through loyal elites and social hierarchies. In the case of environmental hazards as well as for irrigation development in the Indus, science was to masquerade as a politically neutral corpus of thought and practice, only to be deeply complicit with social power, to perpetuate geographies of differential vulnerability to hazards or access to irrigation water.

Not unlike water, power too rarely exists in a pure, unadulterated form. As soon as it leaves the conceptual confines of academia and enters the social world, it bonds with actors, structures, material relations and processes to spawn a bewildering array of spaces, hierarchies, discourses and subjects. To Moore (2018) those power relations get occluded in the institutional and jurisdictional tensions over water, while sectional identities is the only register in which power specifically bonds with water to spawn sub-national conflict. Swyngedouw (1999), on the other hand, in an earlier engagement with the question of science, nation building and the concurrent production of socio-natures, spoke of how modern Spanish nationhood was enacted by Franco's fascist regime through water and irrigation development in Spain. His argument was also about how a seemingly technical enterprise of water development was, in fact, about projecting the powerful gaze of a modernizing fascist regime over Spanish water landscape or waterscape and thereby producing a modern nation state. Technocratic practice and discourse hiding deeply problematic, power-laden notions of human environment and human social relations is not just limited to heroic water engineering; it is also true in the case of mundane water management – for example, see Akhter (2016) on the politics of flow measurement in the Indus – and in the case of hydrological unit classification – for example, see Messerschmid (2015) on the misrepresentation of the Jordan River basin hydrology and political geography by the UN Economic and Social Commission for West Asia (UNESCWA), in favour of one riparian, Israel. They argue that the dominant hydro-social narrative is reproduced by seemingly neutral technical agency, in favour of the basin hegemon. Each of the above examples points towards the imbrication of power relations with science, technical knowledge and then the particularly insidious claims to technical neutrality that work to the disadvantage of some over the others.

In this book I am deeply concerned with unpacking and unveiling the power relations hiding behind the seemingly 'neutral' technocratic gaze. Towards that unpacking, I also seek to destabilize the technocratic notion of stable human–environment relations that are disrupted by freak unplanned accidents. The alternate I will argue through this book is to re-centre the lived experience of subaltern populations, which is characterized by the routinized and normalized uncertainty in human–environment relations, within which conflict over water is embedded. I turn to a framework that privileges uncertainty as normal, while questioning the power geographies driving water conflict in the following section.

Water conflict, in a hazardous world

As a younger academic, I was called upon to teach a first year module in multi-variate statistics. Your scribe was trained in multi-variate statistics, but not enough to teach a whole module in it. But, being a junior academic, as usual, there was little choice in the matter, and the module had to be taught. As one over-prepared for a subject matter in which one was not confident, in a statistics text book I came across the statement that 'average is a predictive model of the data'. And this was the proverbial Eureka moment. Indeed, the 'average' statistic is a predictive model of the data. It predicts that in a normally distributed data 68.2 per cent of the observations will fall within one standard deviation from the mean, and 94.2 per cent will fall within two standard deviations and so on. It is a statistic, which like any other statistic predicts that within a data sample what will be the distribution, and who and what will lie where in a normal distribution. For a social scientist engaged in water and hazards research the conclusion was inescapable. This one predictive model of the data – average – has bonded with the modern societies' social consciousness and world view at the capillary level. Modern society is obsessed with notions of averages and normality. Normality is good; deviance from the mean/average is bad. Hence the entire modern intellectual and planning paradigm has this deep-seated view of mean, average, normal conditions as, well, normal and hence desirable. Modern planning paradigms are, therefore, not unexpectedly, about the question, 'assuming normal conditions persist into a certain time horizon, how do I maximise my return'. In the experience of any researcher like myself, working with traditional and indigenous communities, however, the planning paradigm is invariably informed by the question, 'what if the worst were to happen within a certain time horizon, what do I have to do to be still standing?' It is this disconnect between the human experience of a dynamic uncertain world and the modern imposition of the assumption of normality upon the world that I want to re-centre in this book's conceptual framework – hazardscape.

'Waterscape' is the term often used to discuss the interactions between water, power and sociopolitical dynamics in much of the critical literature on water (e.g. see Menga & Swyngedouw, 2018; Swyngedouw 1999). The term has gained considerable currency, and has encapsulated very well the interactions between technology, state formation and sociopolitical dynamics. The term describes and problematizes the everyday, albeit historically and socially contingent, landscape of water technology and the power relations imbricated in that technology. In their review of the waterscape literature, Karpouzoglou and Vij (2017) list the concept's fluidity as its main strength where, as a 'liminal landscape' (Swyngedouw, 2004), it is always 'fleeting, dynamic and transgressive' (3). Waterscape may provide a robust account of the power relationships producing water geographies; it nevertheless also opens up possibilities for alternate ontologies of water. The above-mentioned strengths of the waterscape concept notwithstanding, its engagement with power relations is distinctly in a political economic mode, and the question of uncertainty and ephemerality is largely an outcome of the relational ontology of water that the concept opens up. To Moore (2018), too, there is a seeming association between 'water shocks', that is, water-related hazards such as droughts and floods, and water conflict, though he wisely doesn't ascribe any causality to hazards. I, however, propose to go a step further and re-centre uncertainty and hazardousness into the political economic imperatives of, say, capitalist accumulation, nation building and empire, or sectional identities that may drive the power relations operative in waterscapes. I posit that the hazardousness of life underpins social relations and is, in fact, integral to human environment relations. Uncertainty is not just the imperative of the post-human ontologies à la Latour (2005), but also an experiential reality of the subaltern, which is gaining increasing urgency with climate change. Hence, I move towards hazardscapes, or hydro-hazardscapes (Mustafa, 2013).

As any researcher in climate change will tell you, it is not just about a warmer planet, or more or less precipitation regionally or locally, but in essence about abandoning the habit of thinking about the world in terms of mean and average conditions. Those averages derived over the last couple of centuries of record keeping are simply not going to hold in a climate change future (Hulme, 2009). What is needed, therefore, is a more hazardous view of the world in such a context – a view anticipating and normalizing extremes, and distanced from the notion of moderate normal conditions. Hazardscape is one such concept, which like waterscape, I argue, is also attentive to scalar politics that are enacted through water, particularly in terms of water as a hazard.

The term 'hazardscape' has been appearing in the literature since the turn of the twenty-first century, but it was first rigorously engaged with by Mustafa (2005) as a derivative of the landscape school within cultural geography. While the notion of landscape also has varied and contested meanings, broadly speaking, according to Henderson (2003), the concept has had four dominant conceptualizations: (1) landscape as *landschaft*, concerned with processes of diffusion and transmission of the material cultural elements (barns, fences, church spires etc.) that constitute folk cultural forms; (2) landscape as social space, that is, the lived space of neighbourhoods, bazaars, shopping malls and so on; (3) the epistemological landscape, or landscape as revealing human practice and ideologies, to be read as a text; and (4) the apocryphal landscape, that is, 'landscape as a way of seeing, especially a way of seeing which relishes the gaze, that asserts power by privileging perspectival vision, which far from being a mere way of seeing, informs the actual, material making of places' (p. 192). Hazardscape combines insights from landscape as social space and apocryphal landscape, but instead of focusing on experiencing and reading landscapes of everyday life, or seeing and asserting power in an aesthetic register, the hazardscape concept instead re-centres the experiential uncertainty and hazardousness of everyday life spaces. It, furthermore, replaces the aesthetic register of the apocryphal landscape with an analytical one – the analytical gaze of power that privileges the expert perspective which, in fact, attempts to materially make and transform places.

Hazardscape was formally defined by Mustafa (2005) as 'simultaneously an analytical way of seeing that asserts power, and a social space where the gaze of power is contested and struggled against to produce the lived reality of hazardous spaces'. The term 'hazards' has unfortunately come to be associated almost exclusively with environmental extremes, for example, floods, tsunamis, earthquakes, droughts and so on. But from relatively early on within hazards literature, and as noted earlier, there has been a profound unease at the roping off of the hazards from the broader continuum of human–environment relations (Kates & Burton, 1986). In fact, it is almost common knowledge within hazards research that more than environmental extremes the normality of 'normal' life and the notion of accidental interruptions of it by unanticipated, unscheduled and dangerous events is the more dangerous fiction, than the events themselves. 'Normal' life is hazardous (Hewitt, 1983).

For a seemingly most mundane topic of water conflict, I maintain that the hazardscape perspective helps us negotiate the complexities of power across spatial scales. The expert problematique that I explained in the previous section spawns realities of water and conflict at different scales. But as I will illustrate through the discussion of the experience of the world purportedly built by the experts, the gaze doesn't go uncontested. The contestation, discursive and material, as will be explained through the case studies is, in fact, as significant in place making as any expert decrees from above. To integrate the hazardousness of the so-called everyday water conflict across scales is not just about normalizing hazards but more importantly about delegitimizing the underlying injustices and power relations of 'normal' life as not just abnormal, but in fact outrageous (Blomley, 2006).

The normally abnormal and hazardous nature of water, I argue, is embedded in experience and even popular culture. A popular 2016 Hindi movie *Mohenjo Daro* tells the fictional story of the Indus River, and the ancient city Mohenjodaro of the Indus valley civilization 5,000 years ago. The movie portrays the Indus River as a mother, life giver and a hazard by virtue of its habit of changing course. The movie, like many Hindi movies, evoked scenes from other popular Hollywood movies like *The Ten Commandments* (1957), *Anaconda* (1997), *Gladiator* (2000) and the apocalyptic *2012*. But, looking beyond the normal oeuvre of a love story and a demigod of a hero, the movie describes how the building of a dam yields gold, but also creates a hazard of the river flowing away. The pugilists in the movie essentially fall on the pro and anti-dam side, echoing the debates around dams in South Asia and the rest of the world more than five millennia after the fictional account in the movie. The good guys, like contemporary environmentalists, pointed to the mortal danger the dam posed to the city – a fear that comes true at the end of the movie – and the bad guys, like contemporary technocrats, to the economic benefits it could bring. Throughout the movie, the sum of all hazards to the city is the river moving away from it, and hence all the rituals, as well as the heroine's chastity, are tied to keeping the river happy, calm and flowing by Mohenjodaro instead of its geopolitical rival Harappa. The hazardscape of the Indus River was the subject matter of an expensive movie to produce – by Indian standards. It told a story which had resonance with the popular conceptions of water – even if the poor artistic and commercial execution left much to be desired.

This book, a little like Mohenjodaro, is about water and conflict at multiple scales and how those scales interact with each other. The fractal nature of the water conflict, that is, self-similar complexity at all scales imbricated with power and complexity, lends itself to a conceptual framework that puts power and uncertainty at the core of inquiry. The question of scale is integral to the concept of hazardscape, insofar as the expert gaze and lived reality and resistance are rarely operational at the same spatial scale. But their effects intersect and produce scales in deeply politicized ways. The production of scale is not just from global to local but is a dynamic that speaks to the interactive co-production of scales between the subaltern and the expert. With these conceptual insights in mind I turn to the geographical context of Pakistan to which is applied the above framework, to apprehend the contestations over and through water.

Pakistan as the most water scarce/vulnerable country, and what does it matter?

Pakistan to run out of water by 2025, scream the newspaper headlines (e.g. see Gulzar, 2018) or that it is one of the most water scarce countries in the world (e.g. see International Monetary Fund, 2015). Meanwhile, water wars loom on the horizon for Pakistan, opine the international observers (Nesbit, 2018). All of these alarmist prognostications speak from or quote the expert view from the national and international scale, through the vocabulary of per capita cubic metres of water. There is very little, if any, acknowledgement that the water system in Pakistan has been modified so profoundly by human action over the millennia that its hydrology can be regarded as more cultural and political than natural (Jacobs & Wescoat, 1994). The physical geography provides the context for the alarmist pronouncements on Pakistan's water, but it does not support the substance of the 'running out of water' thesis, unless the hydrological cycle itself stops functioning for the first time in millions of years. We therefore review in this chapter some of the key features of the physical geography of water in Pakistan, followed by the institutional landscape that provides the context for water conflicts across sub-national scales in Pakistan.

The physical geography of water in Pakistan's Indus Basin

Pakistan is one of the most arid countries in the world and derives almost the totality of its water supply from the Indus River System (see Figure 1.1). The Indus River Basin covers a drainage area of approximately 1,137,819 square kilometres (km^2) between the Tibetan Plateau and the Arabian Sea, the largest portion of which is located in Pakistan (60 per cent), with substantial upstream parts in India and minor ones in Afghanistan and China (Laghari et al., 2012). About 228,694 square kilometres (21 per cent) of the basin area is irrigated, of which 60.9 per cent is located in Pakistan and 37.2 per cent in India. The river system consists of the main stem of the Indus River proper and several tributaries, chief among which are the Jhelum, Chenab, Ravi, Beas and Sutlej Rivers (in descending order from northwest to southeast) and the Kabul River as the only major tributary

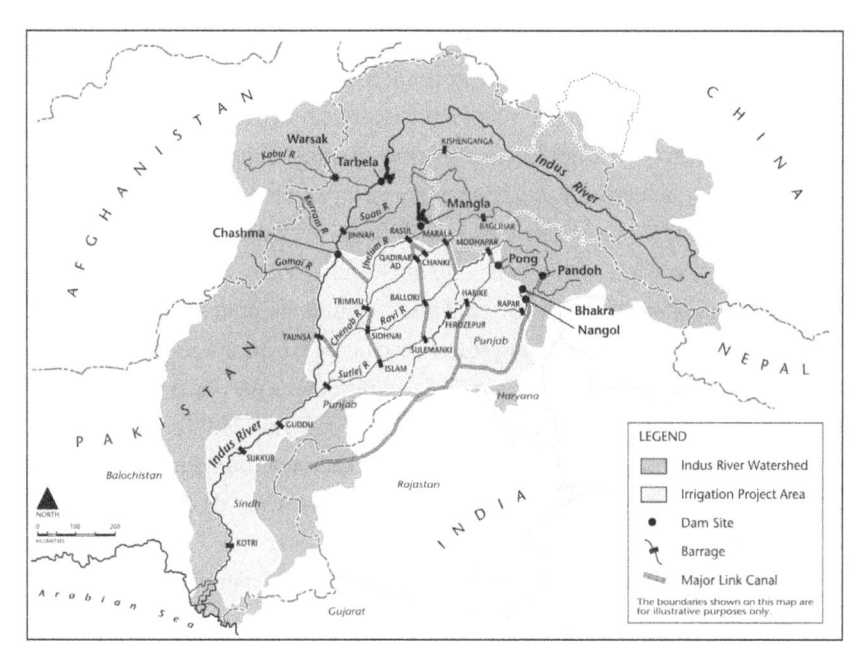

Figure 1.1 The Indus River Basin and its major infrastructure.

flowing in from the West. Glacier melt, snowmelt and rainfall supply the rivers, whose combined annual water flows amount to 207 cubic kilometres (km^3) and enable habitation and agriculture in large parts of the predominantly arid and semi-arid basin. Precipitation ranges from 392 to 461 millimetres (mm) per year and is spatially and temporally highly unevenly distributed, as almost 80 per cent of the rain falls during the summer monsoon season from July through September. In addition to these surface water resources, there are substantial groundwater resources stored in an extensive unconfined aquifer underlying the basin and covering an area of 16 million hectares (ha), 6 million of which are fresh and 10 million are saline (ibid.: 1065). The estimated long-term surface water availability across the basin lies between 194 and 209 million acre feet (MAF), of which 142 MAF are extracted in Pakistan (ibid.). Pakistan's average renewable water availability is about 154 MAF, of which 45 MAF derive from groundwater. Within Pakistan, the province of Khyber Pakhtunkhwa (KP) and the territories of Gilgit-Baltistan and Azad Kashmir are upper riparians, while Punjab is a middle riparian, with Balochistan and Sindh as the lower riparians.

The Indus River is the largest river in the basin and has the unique attribute of having some of the highest silt loads in the world (see Figure 1.2). This is mostly because it drains one of the youngest mountain systems in the world, the

Himalayas, Karakoram and the Hindu Kush mountain ranges. The major but less mentioned consequences of the high silt loads in the Indus system include (1) the rapid silting of any reservoirs on the rivers, (2) in case of flow restrictions or river engineering the aggradation of river channels, therefore accentuating high floods, (3) the need for precision engineering of the canal system and then maintenance of that system for constant slope in the canals to prevent against erosion or aggradation of the canal channels, and (4) the provision of fertile alluvium to the plains through seasonal flooding that maintains the productivity of the land. While the middle and lower reaches of the Indus River System are amenable to the type of surface irrigation developed there, the upper mountainous regions and highlands are largely dependent upon rain-fed agriculture. The lower basin, largely in the Sindh province, faces issues of drainage because of its very gentle slope, something we will elaborate upon when discussing the case study of the fisheries versus farming interests in Chapter 3.

Another important attribute of the Indus Basin in Pakistan to be kept in mind is the groundwater quality. About 64 per cent of the land in the Indus Basin has sweet groundwater and is fit for agricultural and domestic use. The remaining 36 per cent has saline groundwater predominantly in the lower basin and in the interior *doabs* (interfluves) between the main rivers, although pockets of saline groundwater can frequently be found scattered across fresh groundwater zones. In downstream Sindh, more than 80 per cent of the groundwater is saline, making the province almost completely dependent upon surface water supplies for irrigation and in rural areas even for drinking water. In fresh groundwater zones, largely located in Punjab province, groundwater has become the major source of irrigation since the advent of tubewells in the 1960s and meets 80–100 per cent of crop water requirements in certain areas today (Shah, 2007), a facility not available to Sindh by virtue of physical geography. The import of this physical

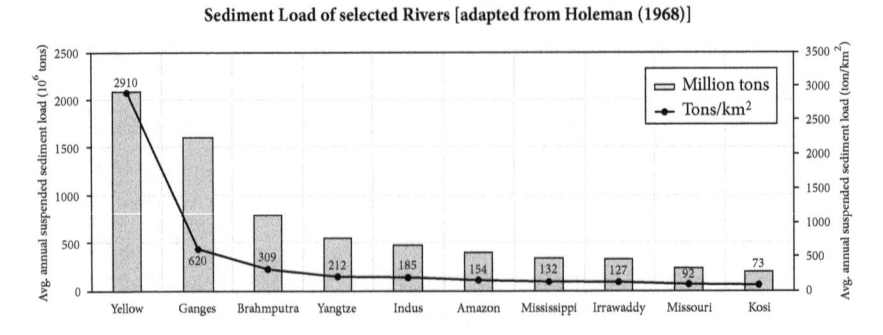

Figure 1.2 Comparative silt loads of major rivers of the world.

geographical reality will become clearer as we discuss the case of inter-provincial water conflict in Chapter 2.

Outside of the Indus Basin, largely in the deserts and highlands of Balochistan, there is a strong dependence upon flood harvesting or spate irrigation. The spate irrigation depends upon channelling seasonal flash floods by check dams to get enough moisture in the land for one or two crops. The system is, however, increasingly under threat, largely because of the high uncertainty inherent in the system, increasingly formalized land property regimes – instead of the flexible land ownership required for such a system – and because of out-migration and a consequent lack of sufficient labour to undertake the work (Van Steenbergen, 1997).

The highlands of Balochistan are also characterized by their dependence upon the Pishin-Lora aquifer for agriculture. The aquifer was historically only passively tapped through the intricate *Karez* irrigation system (see Mustafa & Qazi, 2007). However, not only has the aquifer been depleted at unsustainable rates since the advent of the electric tubewell, but new types of water conflicts are also emerging in its wake. This is something that we shall also be briefly elaborating upon in the chapter on local-level irrigation conflict.

Climate change has to be mentioned in the discussion of the physical features of the Indus Basin as it is beginning to put additional stress on its water resources. This is due in part to the Indus Rivers originating in the Himalayas, with up to 80 per cent of the total average river flows in the basin being fed by snow and glacier melt in the Hindu-Kush-Karakoram mountain range. As a consequence, stream flows in the basin will vary with changes in summer temperature (which affects glacier melt volume) and winter precipitation (which determines seasonal snowmelt run-off volume), and to a lesser degree with variability in monsoon rains in the plains (Yu et al., 2013). In spite of the notorious uncertainties in climate change assessments and predictions for the hydrological regime of the Indus Basin (see Akhtar et al., 2008; Archer et al., 2010; Yu et al., 2013), the possible impacts of climate change on water resources and agricultural livelihoods are of paramount importance for Pakistan's political economy and water security.

With the above physical geographical attributes of water in mind, we now turn to the governance context of water in Pakistan in the following section.

The water governance context

Pakistan is a parliamentary democracy composed of four constitutionally federating provinces. The four provinces, ranked in terms of their population and economic strength, are Punjab, Sindh, KP and Balochistan. In addition to these, there are the Federally Administered Tribal Areas (FATA) (in the process of being amalgamated with KP province), the federally administered territories of Gilgit-Baltistan (which have been given the status of a province with a provincial assembly in 2009, but have no representation and no voting rights at the federal level) and the semi-autonomous region of Azad Jammu and Kashmir, which comprises the Pakistani-administered part of the former princely state of Jammu and Kashmir (Figure 1.1). Water management was designated as a provincial subject in the currently operative constitution of 1973. The federal government of Pakistan has, however, maintained an influential role in water management through the Water and Power Development Authority (WAPDA), which is an autonomous government agency formed in 1958 to oversee the construction and operation of large-scale water storage and infrastructure projects of 'national interest'. The federal government further maintains an active role through the Permanent Indus Commission, which is the main conduit for the government to jointly oversee the administration of the IWT signed with India in 1960. The treaty allocated the entire flows of the three eastern tributaries of the Indus River – the Ravi, Sutlej and Beas – to India. The entire flow of the three western rivers was in turn allocated to Pakistan (see also Akhter, 2015; Alam, 2002; Michel, 1967). The federal government is furthermore an arbiter for inter-provincial water allocations by virtue of the inter-provincial Water Accord of 1991. The Islamabad-based Indus River System Authority (IRSA) administers any conflicts arising out of the accord between the provinces.

At the provincial level, the main agencies responsible for water management are the provincial irrigation departments. Besides the four provincial irrigation departments, each province has a Public Health Engineering Department overseeing domestic water supply in the rural areas and urban Water and Sanitation Agencies (WASA). Karachi's main water agency, unlike other major urban water agencies in the country, is called Karachi Water and Sewerage Board (KWSB), but it has the same line of reporting as the WASAs in other major cities. In addition to the above, the provincial fisheries departments oversee the inland freshwater and marine fisheries within 12 nautical miles of the coast. The 300 nautical miles exclusive economic zone fisheries fall under the jurisdiction of the federal government. The provincial agricultural departments also play a certain role in water management through their on-farm water management units, which provide resources and technical assistance to farmers regarding on-farm water management as well as village watercourse improvement related works.

The formal water management and governance structures outlined earlier are now increasingly being matched at the federal and provincial level by a growing number of vocal civil society organizations, for example, the Pakistan Fisher Folk Forum undertaking advocacy for the mostly poor and highly vulnerable fisher communities in the country. The Fisher Folk Forum is joined by Karachi-based Shirkat Gah to undertake advocacy on gender rights in mangrove and delta environments of southern Pakistan. Farmer Association of Pakistan, a mouthpiece of generally more prosperous farmers that is mostly active in Punjab, is an advocate of large dam construction, especially the Kalabagh Dam (KBD). The Farmer Association on the technocratic side is joined by Pakistan Water Partnership, which is mostly dominated by ex-WAPDA and mostly Punjab and KP ex-irrigation department officials, to again undertake vigorous advocacy in favour of large dams and other water engineering projects. The above cast of big players is also joined by a leftist South Punjab-based Damaan Development Organization, which undertakes advocacy for cultural and ecological rights, with regard to new irrigation development projects in southern Punjab, for example, Chashma Right Bank Irrigation Projects III and Kachi Canal.

Many of the international donor- or corporate-funded NGOs – such as the Hisaar Foundation, Sustainable Development Policy Institute (SDPI), LEAD Pakistan, World Wide Fund for Nature (WWF) and the International Union for Conservation of Nature (IUCN) – are inserting themselves into national water debates – largely on the side of demand management and conservation.

International agencies such as the World Bank and the Asian Development Bank have historically funded large infrastructure projects in Pakistan. However, since the 1990s they too have been supporting initiatives on institutional change, financial sustainability and participatory water management. It is largely thanks to the World Bank's pressure that many canal command areas in Pakistan have been handed over to farmer organizations for administration, which is a move away from the historic bureaucratic management of the irrigation system. Other international actors such as the International Food Policy Research Institute (IFPRI) and the International Centre for Integrated Mountain Development (ICIMOD) have also been active in supporting action research on sustainable water-based livelihoods and basic hydro-meteorological and climate change.

The above physical and institutional context contributes towards Pakistan using almost 97 per cent of its water resources in the agricultural sector with its more than 200 million population using less than 2 per cent of the so-called national water supply for municipal and domestic water supply, with the industrial sector accounting for the remaining 1 per cent (Government of

Pakistan 2018). So volumetrically the spectre of thirsty humans from absolute water scarcity in Pakistan does not matter. But one could argue that using water to grow food is important. There again, within the agricultural water sector, four crops – sugarcane, wheat, cotton and rice – account for 80 per cent of the water use within the agricultural sector. Of the four major water consuming crops two are food crops, but even there, crops such as rice are primarily exported. According to Dalin (2017), Pakistan is the largest exporter of depleted virtual groundwater in the world. The virtual water concept by Allan (2011) is about accounting for the water it takes to produce any commodity from crops to cups of coffee. Pakistan's virtual water exports are largely embedded in its exports of rice, where its water footprint surpasses even that of India, China and the United States. Right in that small detail is how the water scarcity in Pakistan is socially constructed and is embedded in its political economy which drives the conflict over and through water. The national-level picture is not separate from the fact that mostly large farmers at the local scale are engaged in commercial production of crops such as cotton, sugarcane and rice, and in turn influential in driving the political economy of water. The water for rice production also takes away from river ecology upon which previously thriving communities of fisherfolks depended in Pakistan for their livelihoods. This network of dependencies, and water conflicts at and across spatial scales – produced, contested and lived by various societal actors – spawns hydro-hazardscapes that we unpack in this book. In that vein I proceed to outline how my argument is structured in the following section.

How is this book organized

The first issue the book will address will be the inter-provincial water conflict issue in Chapter 2. The chapter will address not only the well-known controversy about large dams especially the KBD between Punjab and Sindh but also water distribution conflicts outside of the KBD between Sindh and Punjab, Punjab and KP, and Balochistan and Sindh. The key problematic for the chapter will be the question of scale where water politics constructed at a national scale tend to conflict with provincial and local-scale politics constructed through water.

In Chapter 3 after describing the historico-geographical context of flood hazard in Pakistan, I will describe inter-provincial conflict over flood and drainage management between Punjab and Sindh and at the sub-provincial scale. I will, in particular, make the case for how the rhetoric and the conflict over floods between the two provinces tend to mirror the nationalist rhetoric around floods in Punjab with regard to India's behaviour during floods. I shall contrast the inter-provincial and local-scale conflict over floods with that over drainage and water pollution. The argument I will try and set up is that power relations are the integrative spine running through conflicts over flood, and irrigation management. Water conflicts are also over local versus national visions of hazardscapes.

In Chapter 4 I will discuss the water conflict at the micro-scale of minor irrigation canals, village water courses and farm gates. The chapter will discuss the physical geography and the historico-institutional context of the Indus Basin irrigation system to set up the context for the case study of water conflicts along two canal commands in central Punjab. The chapter will then outline the institutional and physical structural conditions of the irrigation system that enable water conflict at the local level. The discussion of the conflict will, however, be qualified by the argument that water is often an instrument of conflict at the village watercourse level rather than a driver of it, according to the evidence. The conflict at the canal command level will be contrasted with the conflict over water rights in the Karez irrigation system in Balochistan to make the case that both village watercourse and Karez irrigation conflicts are also linked to the process of scalar production in technocratic hazardscapes.

The penultimate Chapter 5 will shift focus towards water conflict in the rapidly growing urban centres of Pakistan, using the megalopolis Karachi as a case study. After the historico-institutional description of Karachi's water supply, the chapter will highlight two themes of the de facto privatization of domestic water and power politics around water that drive conflict over and through water in the city. The key argument of the chapter is that all the fault lines of ethnicity, language, gender and class intersect in conflicts over and through water in the largest city of Pakistan. The chapter also offers the competition between local neighbourhood scale experiences of water scarcity and conflict versus city and provincial-scale visions of development and hydro-hazardscapes.

The book will conclude by offering some key insights to emerge out of the survey of water and conflict undertaken across geographical scales in Pakistan. Chapter 6 will emphasize the complexity of the forces at play in water and conflict, but will also caution against despair in the face of that complexity. Water conflicts are both discursively and materially driven, and they are an outcome of human institutions and politics instead of some absolute natural scarcity. Conflict over and through water is not because of its absolute physical scarcity in case of Pakistan – there is none. It is because questions of identity, class and development are negotiated at disparate geographical scales from national to regional to local. Chapter 6 will highlight these deeply political and institutional factors driving water conflict and hence will also call for an engagement with water conflict in political terms instead of only technocratic engineering terms. Much of the hegemonic water politics in Pakistan are driven by construction of developmental vision at the national scale while the oppositional politics are embedded in regional- and local-scale identities and hazardscapes. I hope to leave the reader with a keen sense of how such water politics and conflicts are created by humans, and hence can also be undone by the same.

2

Nationalist hazardscapes

The case of inter-provincial water conflict

All those against Kalabagh dam, I say are Traitors! Traitors! Traitors!
(President Lahore Chambers of Commerce and Industry, 03/04/2015)

It is impossible to have a rational debate on KBD or any dam at any forum in Pakistan. The debate on big dams in Pakistan has long since left the shores of reason, and is now in the oceans of religious dogma, and doctrinaire nationalism. If one is a real Muslim (preferably Sunni, Punjabi, male) in Pakistan, one has to be in favour of the most iconic of the large dam projects, that is, the KBD. If one is not, then in the eyes of the proponents of KBD, you have to be a lesser patriot than they are. The above are partially rhetorical flourishes to the very emotive and polarized debate around large dams in Pakistan, but these flourishes are not entirely off the mark either, as the evidence presented in this chapter will illustrate. But why do the emotions spill over on something as mundane as a piece of water infrastructure? We argue that inter-provincial water conflict is best viewed as a hazardscape, where the scalar politics of the hazardscape define the parameters of the debate.

'Pakistan to run out of water by 2025,' screams the headline of one Pakistani newspaper in 2018. This follows the equally dramatic prognostication by the otherwise laconic Pakistan Council for Research in Water Resources (PCRWR), of serious drought and water scarcity in Pakistan by 2025. The same report strongly endorsed construction of large dams to avoid the dramatic projection. Without rehearsing the virtually non-existent scientific basis for the PCRWR's scenario for the entire country, we consider the report, the headlines it provoked and the public anxiety those headlines created as elements of a hazardscape. The analytical minds of PCRWR or WAPDA see an existential threat or hazard of massive drought afflicting the entire country. In the process, the analytical mind doesn't just perceive a hazard but also produces the national scale of that hazard.

The national hazard must then also be mitigated through national projects, such as mega dams. But as we will illustrate in this chapter, the sub-national actors perceive the hazard in very different and less absolutist terms. Their experiential view of the hazard of water scarcity and peril to life and livelihoods is not through abstract projections but rather based upon deeply felt anxieties about history, politics and identity as they intersect with the scalar politics across regions, provinces and nation states.

National emergencies are one very effective means of producing the national scale. Equally efficacious are the developmental visions to do 'nation building'. The large infrastructure infatuation in the water sector is animated not just by fear but also by the hubris of realizing (capitalist) economic growth and development under the assumption of stable environment society relations. Under such an assumption hazards and risks are accidental interruptions of normality, which can be accounted for, and mostly engineered away. In this chapter we will illustrate how the hazardous view of the world, at the heart of the hazardscape concept, animates the strategies of resistance of the less powerful sub-national constituencies, against the 'normalized' accumulative environmental visions of the powerful.

In this chapter I will inevitably review the large dams controversy, as one aspect of the larger upper and lower riparian provinces', namely Sindh and Punjab's, conflict over water. But the chapter will also move beyond the well-known debates, at least in Pakistan, between Sindh and Punjab to also touch upon ongoing controversies between the other provinces on issues such as royalty for hydro-electricity and irrigation water allocation. I will set off by outlining the historical context of the policy debates in the basin, with a particular focus on the history of inter-provincial politics. The chapter will then discuss the inter- and intra-provincial water distribution, and ancillary water-related conflicts between provinces. I will then zero in on the KBD controversy as emblematic of the inter-provincial water politics and how water is imbricated in questions of identity, nationalism and scale inflected notions of development. The chapter will conclude with a discussion of the possible scenarios and conceptual implications of the evidence presented.

As I proceed with the narrative in this chapter, by way of context, it should be borne in mind that the Indus River System has always had rich fisheries particularly in its deltaic environment. Hundreds of thousands of people continue to be dependent upon the fisheries of the river for their protein intake. The high silt loads of the river have facilitated the formation of a complex deltaic environment, and as more water is diverted upstream from the delta for irrigation, and as two of the rivers have been dammed, the silt supply to the delta has been reduced resulting in substantial coastal erosion. The coastal erosion and salt water intrusion are expected to get worse in future climate change sea-level rise scenarios, thereby threatening the livelihoods of the coastal communities, as well as important infrastructure in the largest city of the country, Karachi.

Historical antecedents of inter-provincial water conflict

The signature inter-provincial water conflict between the downstream Sindh and the upstream Punjab province in Pakistan, over water diversions from the Indus River System, has its basis in the development of the irrigation system by the British colonial authorities in late nineteenth and early twentieth centuries. As the irrigation system expanded, primarily in the Punjab, there was growing unease in the lower basin regarding diminished flows into Sindh, which until 1936 was part of the Bombay presidency. As it was separated from the larger presidency, Sindh found its own voice against irrigation-related diversion in the Punjab. The details of the conflict are documented well by Michel (1967); however, the important event to note in pre-independence water debate is the signing of the 1945 accord between Sindh and Punjab, whereby Sindh was awarded 75 per cent of the water of the main stem Indus River and the remaining 25 per cent was awarded to Punjab. For eastern tributaries of the Indus River, 94 per cent of the water was awarded to Punjab and 6 per cent to Sindh. The formula remained in force till 1947 after which the water allocations were largely undertaken between the provinces by the federal government on an ad hoc basis. Those allocations were largely perceived by Sindh to have favoured Punjab.

The second most important event in inter-provincial disputes was the signing of the IWT in 1960 as discussed in Chapter 1. The immediately relevant fact with regard to inter-provincial water disputes was Sindh's perception that Punjab had bartered away its water to India and was now compensating itself by diverting water from Sindh's share, as the following quote by a Sindhi nationalist activist illustrates:

> According to IWT Pakistan sold 3 of its rivers to India and they got a price for it. At the time the headworks were in India where the gauges were installed. The water that would come from India was 3MAF and 2 of the MAF would go past Panjnad. Punjab in order to make up for the shortfall of water made link canals from the Sindhu (Indus) River without the consent of Sindh – they started stealing Sindh's water. And then they took loans from international agencies to build Tarbela Dam. And that water went to Punjab's benefit as well. (Vice President, Jeah Sindh Qaumi Movement, 23/04/15)

The emotions that the inter-provincial water question provokes in Sindh are somewhat justified by the physical geography of the basin. The monsoon rains in Pakistan are strongest in the northeast of the country, and the influence of those rains keeps getting weaker as one moves towards the south and the west of the country. Consequently, in the south and the west of the country, the dependence of the societies on the Indus River water is more intense than in the north and the northeast.

In addition, western most Indus River's bed has the highest elevation of the six rivers in the system. This simple geographical fact has been very crucial in the water management of the basin, especially in the aftermath of the IWT. As the water from the three eastern tributaries was awarded to India, the World Bank along with a consortium of Western donors funded the construction of link canals to transfer the water from the three western rivers to the eastern rivers as they entered Pakistan to compensate for the water India was going to withhold for its use. The gravity-based link canals were largely feasible because of this simple geographical reality, and the operation of those link canals has become part of the controversy between Sindh and Punjab, particularly of the Chasma-Jehlum link canal (Figure 1.1). The construction of the link canals was deemed to be the evidence of Punjab's intent to appropriate what Sindh perceived to be its water. To many in the Pakistani bureaucracy the signing of the IWT and subsequent infrastructural development works was an exercise in producing a national scale. A sovereign country engaging in international diplomacy and then benefiting from technology transfer to build the basis for a modern developmental future was the aspirational outcome of the production of this national scale (Akhter, 2015). But to the Sindhi activists, like the one quoted a few paragraphs earlier, all the oppositional engagement is about destabilizing the national and instead reasserting the primacy of the regional and provincial scale. The scalar disconnect in this hazardscape is quite stark indeed.

The third important landmark agreement was the 1991 Inter Provincial Water Accord (the 'Accord' from here on), which resulted in the formation of the IRSA for the implementation of the Accord. The water allocations and the surplus water allocations in the Accord are given in Table 2.1. The Accord provides for a minimum of 10 MAF of water to escape downstream into the delta, even though historically on average there have been about 35 MAF of water escaping into the delta (Afzal, 1995). This was also one of the first instances of the post-independence Pakistani state formally conceding some hydropolitical space to the sub-national and provincial scale. Water management is a provincial subject under the 1973 constitution of Pakistan, but inter-provincial water management

Table 2.1 Agreed Apportionment of Water according to the Inter-provincial Water Accord 1991 (figures in MAF)

Provinces	Kharif	Rabi	Total
Punjab	37.07	18.87	55.94
Sindh*	33.94	14.82	48.76
(a) Khyber Pakhtunkhwa	3.48	2.30	5.78
(b) Civil Canal**	1.80	1.20	3.00
Balochistan	2.85	1.02	3.87
Total	77.34	37.01	114.35

Surplus river supplies (including flood supplies and future storages)

Punjab	Sindh	Balochistan	KP	Total
37%	37%	12%	14%	100%

*Including already sanctioned urban and industrial uses of Metropolitan Karachi
** Ungauged civil canals above the rim stations
Source: IRSA

is not (Alam, 2019). But even this concession is founded upon an average/normal perspective on water as per Table 2.1.

Much of the discussion in section 'Agreements, Allocations and Royalties in Inter-Provincial Water Politics' with regard to the reality of inter-provincial water allocation between Sindh and Punjab will be with reference to the provisions of this accord and its legitimacy and meaning in the eyes of the two provinces in particular, and all the signatories in general. For now, in the following section we will outline the history and contemporary reality of the wider inter-provincial relations in Pakistan, to set up the political context within which the inter-provincial water disputes must be understood.

Politico-historic context of inter-provincial relations in Pakistan

Since the independence of Pakistan in 1947, inter-provincial relations have been marked by contestations on political, economic and cultural issues. From the very outset, the country emerged as a centralized state run by an unelected political elite and bureaucracy. The unrepresentative state institutions, and the lack of mechanisms to articulate political demands – the first general elections took place in 1970 – left provinces and ethnic groups living within them disenfranchised, creating an enabling environment for adversarial ethno-nationalist politics.

Rehearsing the causes of centralization in Pakistan might be digressive here, and it is nevertheless important to mention that inter-provincial mistrust and grievances of provinces against the centre and each other were the outcome of denial of political agency to the provinces. Punjab, Sindh, KP and Balochistan had had their distinct political and cultural identity even during the British colonial time, but all four provinces were merged into one unit of West Pakistan in 1954, depriving them of their autonomous status as of August 1947 (e.g. see Khan, 2009).

In the first two decades, the centre justified one unit policy by playing upon the fears that with 55 per cent population of the country, (Bengal) East Pakistan would dominate the rest of Pakistan in a decentralized federal system. However, the policy of a strong centre did not work. The Bengali nationalist movement, which eventually led to the separation of Bengal from Pakistan in 1971, drew its legitimacy from unjust economic policies of the state and a denial of cultural identity to Bengalis.

Being the biggest province in terms of population after the secession of Bangladesh in 1971, Punjab dominated the politics of the country, which can also be attributed to the fact that the leadership of bureaucracy and military – the two institutions that controlled the reins of the state – by and large came from the province. Today the ethno-nationalist elements have similar bitterness and a sense of grievance against the Punjabi elites, for example,

> We (Sindhis) have been a nation for a very, very long time, much before anybody had ever heard about Pakistan. We have been Sindhis for thousands of years, and we have been living as an independent nation for the longest. . . . This (Pakistan) is an unnatural country. In the world there are two countries made in the name of religious, Israel and Pakistan. In the world there is no conception of religion-based countries. Religion based countries are unnatural, anachronistic and unscientific. Why do we have 40 Islamic countries? Why aren't they together. (Vice President, Jeah Sindh Qaumi Mahaz, 23/04/15)

Comparable sentiments are echoed by ethnicities within Punjab as well, for example, the Seraiki intellectuals in southern Punjab:

> Have you ever met a person who was a Pakistani before 1947? Nations of Pakistan are a very relevant issue. Before Pakistan it was Sindh, Balochistan, Pakhtunkhwa and so on. We were the people who were here before Pakistan. (A Seraiki nationalist leader 28/04/2015)

Sindhi and Pashtun ethno-nationalisms started gaining currency because of a sense of deprivation in Pashtuns and Sindhis. GM Syed, an iconic Sindhi political leader, who was part of the Muslim League and who supported the Pakistan Movement, became disillusioned with Pakistan and started voicing secessionist political views, arguing that Pakistani state had become a Punjabi state where the interests of the latter trumped the rights of the other provinces.

Balochistan is a different story altogether. The province has witnessed five waves of secessionist nationalist movements and insurgencies and as many military operations since 1947. Some political groups in Balochistan, where the anti-Punjab sentiment runs deep, have never accepted Pakistan, saying that Kalat state – an independent Baloch state that had an independent pact with the British Crown in London and not in Delhi – was forcibly annexed by Pakistan.

Within the above-mentioned atmosphere, the founding party of Pakistan and the party of Mr Muhammad Ali Jinnah, the All India Muslim League, later Pakistan Muslim League (PML), was appropriated by the politico-military elites to become a party of the establishment and of the political right. That legacy continues with all the factions of the PML. The party's core support is in the Punjab with regional factions catering to conservative or opportunist constituencies in all the four provinces. Pakistan People's Party (PPP), the left of centre party in Pakistan initially had its core support in the Punjab but over a period of time it has ceded much of its political ground to ethno-nationalist parties and after the elections in 2013 has been largely left as a party of rural Sindh. Besides these two mainstream parties there is also the right wing Pakistan Tehrik-e-Insaf (PTI) appealing to the middle class in mostly urban areas of Punjab, KP and Sindh. While PPP has maintained a studied ambivalence on inter-provincial water issues, PML has been a vocal proponent of Punjab's view on inter-provincial water issues. PTI has also remained ambivalent but from conversations with their representatives, it appeared that they were keen to undertake large water development projects.

Almost all the ethno-nationalist parties, for example, Awami National Party (ANP), Jeah Sindh Qaumi Mahaz, Pakistan Taraqqi Pasand Party, Balochistan National Party, Pakhtunkhwa Awami Milli Party (PAMP) and others, are on the left of the political spectrum, against large dam projects and in favour of the positions of the smaller provinces. All right wing parties, for example, Jamaat-e-Islami and Jamiat-e-Ulema-e-Pakistan, are generally in favour of the central government, and in agreement with the Punjabi arguments in favour of large dams. Mutahida Qaumi Movement (MQM), the centre left party of the former migrants from India, which dominates urban Sindh, is against large dam projects out of solidarity with the province that they are in, though their position could change if their relationship with the larger Sindhi polity undergoes a major change.

In the absence of democratic and federal institutional frameworks, inter-provincial conflicts and ethno-nationalism were left to metastasize. The above-mentioned political parties have been the conduit through which many of the inter-provincial tensions are articulated. Things have, however, started changing with the passage of Eighteenth Amendment – a big step towards substantive federalism and democracy – which abolished the concurrent list, giving more autonomy and resources to provinces.[1] However, the mistrust between the provinces is deep-seated and still unaddressed, as some of the interviews of political party representatives in the following section will illustrate.

Agreements, allocations and royalties in inter-provincial water politics

As the foregoing discussion on inter-provincial and intra-provincial politics illustrates, the relationship between the federating units and between ethnicities within those federating units has been fraught with tensions between secularist decentralizing tendencies of ethno-nationalist movements and the centralizing tendencies of the military-bureaucratic elites populated by northern Punjabis, Mohajirs (migrants from India since 1947) and, to a lesser extent, Pashtuns from KP. Those tensions are also manifest most poignantly with regard to the large dam controversy (which I will get to in the next section), but also less dramatically with reference to everyday allocations of water as per the 1991 Water Accord. The most acrimonious debate with regard to water allocations is between Sindh and Punjab. Many contemporary Sindhi nationalists are not enthralled of the 1991 Accord primarily because it was signed by a propped-up regime,[2] which many argued was illegitimate at the time, by Jam Sadiq Ali, the then chief minister of Sindh, for example,

> We disown the 1991 Accord because it was signed by Jam Sadiq Ali, who did not have the popular mandate. It was forcibly signed. Even then it is not being followed. The 1945 accord should be the basis of future negotiations between Sindh and Punjab, and Sindh should be compensated for the water stolen in the interim. (Ayaz Latif Palijo, President, Qaumi Awami Tehreek, 23/04/15)

---

[1] The Eighteenth Amendment to the constitution of Pakistan was passed in 2010. As per the amendment, almost all the powers of the federal government except foreign relations, defence, communication and finance were transferred to the provinces.

[2] Jam Sadiq Ali's coalition government was propped by the central government and country's intelligence agencies as a bulwark against the return of the Pakistan People's Party in the province. Sindh is the heartland of People's Party, and at the time, the Pakistan Muslim League government and the military thought that Pakistan's interests would be best served if the same party or its allies ruled all the provinces (Afzal, 1995).

While this sentiment about the illegitimacy of the 1991 Accord was repeated by many other Sindhi nationalists – among the mainstream there was a realism about its reality and a desire for it to succeed. But its perceived non-implementation and violation by Punjab was considered evidence of Punjab's bad faith vis-à-vis Sindh, for example,

> We did not accept it [1991 accord] but there is no other formula for the division of water. So now it should be honoured. When this Accord was signed, Sindh was against it because Jam Sadiq Ali was not a popular CM of Sindh. And there were lots of objections presented. But since it was based on 10 day cycle, and there was a formula for surplus water between dam fillers, the formula was agreed. We wanted that 1945 agreement should be the basis for the Accord [but it is too late for that]. (Qadir Magsi, President, Sindh Taraqi Pasand Party, 23/04/15)

The issue of the 1991 Accord was often linked to the international-scale signing off of the three eastern rivers to India under the IWT. Therefore, in the minds of the Sindhi political leadership, the *meso* inter-provincial scale was intimately linked with the marco-scale picture of international treaty with India, for example,

> My party rejects the 1991 Water Accord. We believe that the basis of the negotiation should be the 1945 treaty. It is their fault that they signed off their rivers to India, why should we pay for their mistakes because of which the 1945 accord cannot be resurrected? (Vice President, Jeah Sindh Qaumi Movement, 23/04/15)

The main bone of contention on part of the Sindhis with regard to the implementation of the 1991 Water Accord is twofold: (1) that Punjab instead of delivering the requisite water to Sindh satisfies its historic demand, particularly during times of high water demand before releasing water for Sindh, and (2) that the link canals particularly Chashma-Jehlum Link and Trimmu-Panjnad Link Canals are supposed to be flood canals to be only operated with the consent of the chief minister of Sindh, but are operated somewhat arbitrarily on orders of the Punjabi-dominated IRSA. Of the two main objections, there is strong evidence to suggest that the first one is quite true, both in terms of quantities of water delivered and more importantly the timings of the water delivery:

> Sometimes we have ways of taking more water than our allocations in Punjab because they will waste water and it will all go to the sea. So, at night or otherwise, we will take more water. For example, if 28,000 cusecs is coming through and we have to pass on 25,000 cusecs we say we only have 22,000 coming. But we

do that too with the knowledge of the Sindhi Executive Engineer (Xen) posted at the barrage. We just ask him not to come out of his quarters at the time of the diversion – and he doesn't. (Superintending Engineer, Punjab Irrigation Department (PID), 27/04/15)

The Accord specifically provides for water demand by the provinces to be decided based on the average water demand in the preceding ten days. The surpluses and shortages are to be shared on a pro-rata basis with reference to absolute proportional share of the provinces (PILDAT, 2011). Punjab contended back in 1994 that since KBD had not been constructed, it was not possible for Punjab to fulfil its obligations to Sindh and that the water allocations will be as per the post-Tarbela 1976–81 period averages in the Punjab. The government of Punjab's cabinet's decision to that effect was annulled by the Law Division of the Government of Pakistan in 2002; nevertheless, unofficially Punjab continues to behave as if the 1976–81 averages are the basis for water allocation. In fact, Punjab tends to interpret Clause 6 of the Accord providing for construction of additional storage in the basin as a reason to build KBD. Sindh, on the other hand, interprets that to mean local-level projects and not a national-level mega project like the KBD, for example,

> 'Whoever has the stick, owns the buffalo,' Punjab always uses the words wrong. The Accord just says that the provinces can build schemes for their own use. Kalabagh is a separate issue. (Qadir Magsi, President, Sindh Taraqi Pasand Party, 23/04/15)

There is an important statistical trick at play here. Punjab while ignoring the ten-day allocation principle often protests that it ends up fulfilling Sindh's annual quota, as one Xen from Sindh Irrigation Department explained:

> You are absolutely right about average flows. If Punjab insists that it on average delivers the requisite volume of water, it would be correct. But in the past 15 years for 12 of those years it would deliver a lot less than the average flow. But when you add in the flood flows from say 2010 and 11 you will of course get the requisite average for those 15 years maybe even more. So, I think we [in Sindh] are really hurting ourselves by sticking to the average based policy in case of water. (Xen Sindh Irrigation Department, 26/03/15)

Sindh's position is further elaborated in Figure 2.1, which lists the annual flows of water downstream of Kotri Barrage, the last structure on the Indus before it enters the Arabian Sea.

As Figure 2.1 illustrates some years the flow can be quite low, and being that riverine communities depend upon the waters of the Sindh even for drinking

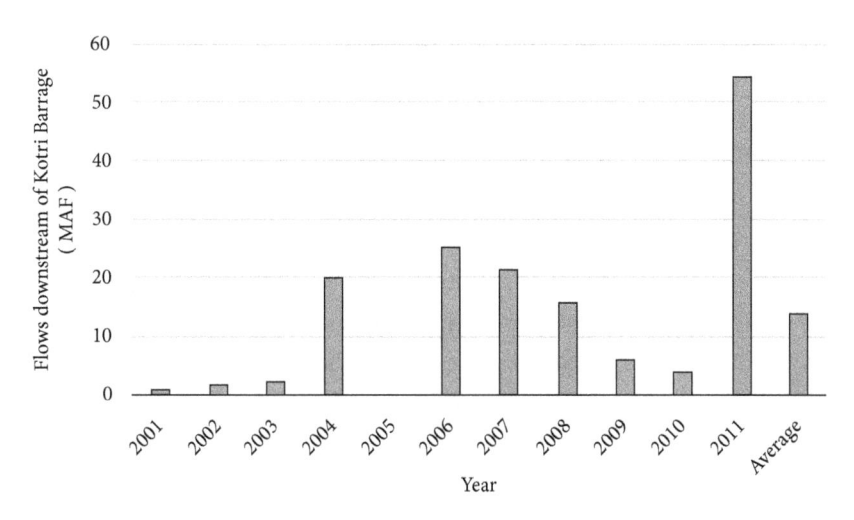

Figure 2.1 Flows in MAF downstream of Kotri (PILDAT, 2011).

water, Sindh's position about variability apart from the averages is borne out. As a member of IRSA pointed out,

> There are a million shortcomings in us Sindhis, I concede. We may be lazy and not much for working too hard, yet we don't deserve to die of thirst. . . . At least amongst us you will not find a single blasphemy case, unlike Punjab. . . . All the floods have recharged the groundwater in Sindh. So, they were a great blessing. (A Sindhi Federal Official, 11/09/15)

Elaborating further on the timing issue he said:

> One to two years ago Sindh got water quite late. The Chief Minister Sahib called me and asked me what is the story with the delay in delivery of water. I said I have released water as per the accord [from Tarbela and Mangla], but it takes the water 20 days to get to Kotri which causes problems. And it is during that time that all hell breaks loose in Sindh. (A Sindhi Federal Official, 11/09/15)

The issue is not just of timing till the water reaches Sindh, but also of interseasonal water flows downstream of Kotri. Often times outside of the Monsoon season there are virtually no flows downstream of Kotri Barrage. The story of timing and average flows gets similarly repeated by Sindh with Balochistan, even by Sindh's own admission:

> In my division we frequently have 50 Baloch sardars (tribal leaders) show up with 200 armed men each, and shut down every regulator along the Kirthar canal per force, to more than satisfy Balochistan's demand for water. Here I show you the communication I have with the Balochistan Xen on water. He is getting

more than his share of the flow into his jurisdiction – but you see I am pulling the same trick as Punjab – I am delivering more water in Feb. when I don't need it [I probably will hold back if there is scarcity] so in the long run Sindh is in the clear. (Xen, Sindh Irrigation Department, 26/03/15)

According to the 1991 Accord, Sindh and Punjab are supposed to bear shortages to provide for the relatively minor shares of Balochistan and KP (Table 2.1). But even those minor shares within the context of a canal division, especially when there is high water demand, can mean a significant problem at the local level. In fact, in 2009 armed tribal farmers took over the Xen Sukkur Barrage's office, along with the chief engineer, irrigation, Balochistan Irrigation Department. The chief engineer of Balochistan Irrigation was extended the courtesy of operating the Sukkur Barrage to see for himself how he might be able to supply the requisite water at the sowing time, so that it could reach the tail ends of the Kirthar Canal in Balochistan. The Kirthar Canal branches off from the North West (NW) Canal that starts from Sukkur Barrage. The chief engineer noted in a letter that he was convinced that the discharge at Sukkur Barrage had to be about 130,000 cusecs in order for the NW canal to be able to supply the tail end of the Kirthar Canal. The Xen Sukkur Barrage, however, noted to the researchers that he needed at least 100,000 to be able to satisfy Balochistan's demand.

Besides the frequent forays of angry and armed Baloch farmers into Sindh to ensure water supply to themselves downstream, there is frequent communication between irrigation officials on both sides by text messages. The tenor of the communication can be professionally confrontational, as I witnessed looking over the transcripts of the communication. For example, in a communication transcript that we saw, the Xen in Balochistan wanted the regulator for the canal to Kirthar to be closed immediately, because there was too much water in the Kirthar Canal. The Sindhi Xen protested that he could not because canal on his side would breach if he were to close it immediately. In another instance the Balochistan Irrigation Department Xen curtly reminded the Xen in Shadadkot Sindh that Balochistan has a water right that needs to be respected and fulfilled.

The issue of water distribution has a relatively less well-known intra-provincial dimension as well, particularly when it comes to southern Punjab where the Saraiki-speaking people have a long-standing demand for provincial status. They too share the other ethnicities' antipathy against northern Punjabi domination, for example,

When Sindhis make noise about not getting enough water, the IRSA sits down and gives away Saraiki belt's share to Sindh instead of making a sacrifice by itself.

At least they should acknowledge that Saraikis make the sacrifice, not Punjab. Punjab should give its own water, why give our water. That is because nobody listens to us and our movement is not strong enough that we can block the road or kills few people Altaf Bhai (MQM leader) style. (Abdus Sattar Phaeen, Seraiki Nationalist Activist, 27/04/2015)

The Saraiki nationalist are resentful about the allocation of the Sutluj River to India and the consequent human and ecological consequences of it in their homeland. They are also resentful of their lack of formal voice within the Pakistani federation where they deemed the (then) ruling PML (N) as a northern Punjabi ethno-nationalist party.

A central character in the story of inter-provincial water conflict is the 'average flow'. Average flow is the locus of compliance and violation of the 1992 Accord. The expert national and provincial level view is grounded in normalizing the inherently temporal and spatial variation in water flows within the Indus Basin. The activist view, too, creates a provincial and a regional scale where the normality of the expert mind is contested and politicized. The pretence of an apolitical, technocratic water management regime is the fiction underlying expert-level inter-provincial water conflict. Within that universe, occluded are the activist voices pointing to regional-scale hazardscapes, where politics, identity and vulnerability are constructed and contested. The regional and provincial hazardscapes are not just about potential and real water scarcity but also about claim making on the national scale, and that too through hydrological infrastructure like large dams, for example, the case of KP and its claim to royalties from the Tarbela Dam within its territory.

In Pakistan KP is an upper riparian rich in water resources and power generation potential. Its main grievance with the federation is that it be paid the royalties due to it from Tarbela Dam and other electricity generation projects from which it supplies power to the Punjab. The following quote encapsulates KP's leadership and possibly public's position on the water issue:

I am from Swabi and there were people who were displaced from Tarbela, but they still haven't been compensated. Electricity is being supplied all over the country, but then the same electricity when supplied to KP is supplied for twice the price. When Punjab sells agricultural products to us, it charges us the market price. But when it comes to electricity everybody is charged the same price. The government when it sells to KP is selling at twice the cost price. So, we are saying where are the profits and when do we see the money? (Senator Sitara Ayaz from KP, 31/07/2015)

The legitimacy or reasonableness of KP's claim over royalties or demand for lower price electricity is beside the point. The point is that it is a demand that is within the agenda of the political leadership of the province and the consciousness of the people. In fact, as a former minister from the People's Party said, at the time of the National Finance Award negotiations, KP would not even join the discussions until the federal government reimbursed it for the royalties it had claimed from hydro-electricity. The only way KP can realize value from its water share under the Accord is either through development of its irrigation network on limited viable land or through selling water or realizing royalties from electricity generation.

The sum of the above discussion is that inter-provincial water allocation is a perpetual source of conflict between the water bureaucracies of the provinces. The upper riparians invariably do what they can to fulfil their immediate needs, before they turn to addressing the needs of the lower riparian. While the inter-provincial water thefts and conflict are leveraged by politicians to gain political mileage, the problem is related more to (1) the political imperative for every province of satisfying the water demand of local farmers in the first instance; (2) the facts of geography; and (3) the conceptual models and frameworks used in water management, for example, average monthly or annual flows – than any exceptional ill will on part of one province or another. Even an upper riparian like KP tends to insert itself into the water discussion by its claim to unpaid royalties to it for Tarbela Dam and electricity generation on its territory that gets supplied to the rest of the country. It is the dam discussion that is the most emotive part of the inter-provincial water conflict in Pakistan, and perhaps much less driven by structural constraints and more by engineering hubris, and possibly even ethno-nationalist pride than anything else. Large dam projects also frame the national-scale hydro-hazardscapes and the contestations inherent in those, as I shall illustrate in the following section.

Kalabagh or death/death or Kalabagh: Competing narratives of inter-provincial water conflict in Pakistan

KBD has emerged as the lightening rod issue in inter-provincial water debates. The dam is the unfinished business of the Indus Basin development projects funded by international donors in the aftermath of the IWT in the 1960s and 1970s. Part of the reason for its long shelf life, as opposed to coming to fruition, may have to do with the dissolution of the One Unit in 1970 whereby Sindh regained its voice as a federating unit. The second may have been the focus on the Tarbela Dam back in early 1970s because of the considerable number of engineering problems it had in the beginning. In fact, initially KBD was proposed as a priority instead of Tarbela by the World Bank, but Pakistan at the time chose to go with the Tarbela Dam. Furthermore, the change in international finance institutions' priorities away from mega dam financing proved another barrier in its initiation in the late 1970s and 1980s. Over the more than fifty years of its existence on paper, KBD, a proposed dam about 300 kilometres downstream from Tarbela on the main stem Indus River, has become an emblematic project that mediates different notions of the meaning of the Pakistani state, development, ethno-nationalism and the rights of the smaller ethnicities in Pakistan.

Kalabagh is not the only dam planned in Pakistan. There are others, for example, Bhasha, Dasu and Katzarah Dams upstream of Tarbela Dam to which all the federating units have agreed to. The foundation stone for Bhasha Dam in Diamer district of Gilgit-Baltistan was laid in 2008, but no substantial progress has been made for its construction. Part of the problem is arranging financing, because the dam is located in a disputed territory, and second, its proposed site also happens to be in one of the most seismically active zones in the world. Nevertheless, it is the KBD that has captured the imagination of the Punjabi and Sindhi leadership, and the public. Figure 2.2 shows a banner on the main Mall Road in Lahore, conveying the depth of feeling that the Punjabi populace, or interests there in, bring to the construction of the dam.

Figure 2.2 A poster along Mall Road, Lahore, urging Pakistan's popular chief of the army staff (COAS), General Raheel Sharif, to undertake the construction of Kalabagh Dam.

The poster has photographs of the then Pakistan chief of the army staff (COAS), director general Inter Services Public Relations (ISPR) and a couple of Army Corps Commanders. It literally reads as follows from top to bottom:

> Step forward [presumably the COAS] all of Pakistan is with you. Oh son of Pakistan, leader of the families of martyrs, and the leader of the armed forces of Pakistan. General Raheel Sharif we demand of you to build the Kalabagh Dam. [We demand] The initiation of [another] zard-e-azb for the completion of Kalabagh dam and other energy projects (Zarb-e-Azb is the name of the ongoing military operation against Tehrik-e-Taliban Pakistan in North Waziristan). For a permanent solution to India's hydro-offensive ongoing since 1947, to liberate every child and every corner of Pakistan from debt, mortgage and under-development – we demand another zarb-e-azb [for Kalabagh].

The forceful message of the poster along with many others right next to it is somewhat representative of the depth of feeling illustrated by the quote at the beginning of this chapter by the president of the Lahore Chamber of Commerce and Industry. It is also illustrative of the nationalist hazardscape where national-level water scarcity and India's cutting off of water to Pakistan are the signature hazards at the national scale.

For most, dam is not something to get too emotional about. Reasonable people can disagree about the necessity of this or that infrastructure project, but not in Punjab or Sindh, and not about KBD. As a high-ranking water-related federal official said,

> There is mistrust between the provinces, when one brother feels grievances against another brother. I asked one person in Sindh, who is an opponent of the dams, why don't you complain about Indian dam building? There are some politics going on here. If you build the dam first, your rights will be developed. So, there is a conspiracy to prevent Pakistan from building dams so that India can develop its rights. (A federal official, 17/04/2015)

So, presumably as per the above, anyone opposing KBD is conspiring with India to prevent Pakistan from developing or perfecting its water right. This is an oft-repeated accusation, particularly by the Punjabi elites, as well as right wing political parties. Most recently, in the news coverage of a report by the International Water Management Institute (IWMI), which described Indus Basin as a closed system where there is no additional water for further storage, the paper cryptically evokes the Hindu-sounding name of the author of the report to further legitimize the supposedly neutral IRSA's objection to the findings of the report (Kiani, 2015). If one were to accept the findings of the report by IWMI, the entire case for KBD would disappear, and in the eyes of some Pakistani readers, especially the ones in favour of KBD, a Hindu analysts' case against KBD would be a clear affirmation of a conspiracy to undermine Pakistan's development. The KBD in Pakistan's water engineers', and the Punjabi populace's, imagination is indispensable for their vision of development, which was articulated very well by Mr Shams-ul-Mulk, former chairman WAPDA and chief minister of KP:

> If you look at the world, there are three main powers, India, China and America. America has built 6500 dams, China has built 22,000 dams in the past 50 years, India has constructed 4,500 dams so far and 650 are under construction. There is no need to add anything to the above, the data speaks for itself. (Shams-ul-Mulk, 9/12/2015)

In the above parlance there is only one developmental trajectory that Pakistan or, for that matter, the entire world can follow, that of India, China and the United States. The way to fulfil that destiny is through dams, starting with KBD. Like most educated Pakistanis, Mr Mulq may or may not believe in God, but his belief in modernist Western-style development is absolute. KBD is key to realizing that type of development.

While the emotions surrounding KBD in Punjab might border on the hysterical, they are no more rational or cool headed in the other main party to the debate, Sindh. The Sindhi nationalist leadership, however, is much more aware of the technical details of the water situation in Pakistan than their Punjabi counterparts. Part of it may very well be the function of being the lower riparian, and being totally dependent upon surface water supplies, because of mostly saline groundwater. The Sindhi nationalist political mood, much like the other ethno-nationalist political parties in other provinces and in the Saraiki belt, is largely left wing. Consequently, issues of identity, ecology and equity have much greater resonance with them, than the nation state scale development paradigm that the largely right wing dominant political ethos of Punjab might allow. In fact, remarkably at the national level the positions on KBD are neatly split along the right and left wing ideological divide in Pakistan, for example,

> I am not an expert in the field but we need both big dams and smaller dams with 'run of the river' projects wherever feasible. Chinese and Indian examples are instructive. . . . Storage has been blocked by controversy over Kalabagh dam. It was a very desirable project. Unfortunately, the political leadership in the country including WAPDA did not act at the right time and in the right manner. I am not hunting for conspiracy theories, but I would not rule out foreign finger

in creating antipathy against the Kalabagh dam.... I do not agree with Sindhi narrative of distrust. There is some truth in their argument, but I think it is very tainted and I do not agree with the narrative. Indus River belongs to everybody even if it is named after the lower riparian. We need a model for resolving these issues. Sindhi people believe that it is at their cost and there is some truth in it. Balochistan, though not part of the Indus system, is also a stakeholder. It too should be involved in decision-making. (A leader of Jamaat-e-Islami, 13/07/15)

The narrative on part of the left wing ethno-nationalists in Sindh is illustrated by the following quote:

Efficient management of water is the fundamental question in Pakistan. Tell us what is the point of building dams? Do you want to grow food? Electricity? Or do you want to oppress Sindh and control their water for eternity? If you want to generate electricity, what have you done since Bhutto to generate electricity? Why don't you go for other options? Do all countries go for hydro? Why not have 'run of the river'? Is anybody thinking of the country, reducing water wastage, do laser levelling and why not do that all over Sindh? Why not have efficient use? Since the British the canals have not been improved. (Qadir Magsi, President Sindh Taraqi Pasand Party, 23/04/2015)

The preceding two quotes illustrate the conflicting national versus provincial- and regional-level scalar production through water. The KBD is pivotal to materially producing a national scale with its stream of benefits flowing from KBD. Equally at the provincial scale KBD will undermine the provincial scale favoured by the ethno-nationalist, with its stream of hazards flowing entirely to the sub-national scale. The issue is indeed the lack of trust between Sindh and Punjab, and that much was admitted by all the political parties on the right and left of the political spectrum. But the scalar production is what underlies that lack of trust. Evidently Punjab's behaviour as documented in the preceding section does not quite inspire confidence on part of the Sindhis. The key objection to Kalabagh in the Sindhi mindset is about the designed off take of canals both on the right and the left banks of the Indus, presumably to irrigate lands in Southern KP and Punjab. But the Punjabi position typically tends to overlook those critical features and emphasizes the energy potential of the dam. In the words of a Sindhi engineer:

It [KBD] has become so emotive that sometimes when I meet Punjabi friends, they say things about KBD that are simply irrational and downright provocative. They say that there would not have been destruction in Badin and Tharparkar, if there was KBD. They say that they are only building it for Sindh. Why don't they just be straight that it is for the benefit of Punjab, and Sindh may benefit too.

Who are they trying to fool? At least I could buy that. Why won't they build the ones that they have already authorised? Why won't they touch Bhasha [dam], which has already been authorised for 10 years? They have just become obsessed with KBD. . . . If they [the Punjabi water bureaucracy] had not lied, it would have been built in 1985. 'In Murray Darling they have 1000 day storage' they say, try doing it in the Indus [referring to the size of the river and how it would be simply not feasible to have infra-structure big enough to store a 1000 days of Indus' flow]. They are doing a massive modelling exercise for the Indus and the consultants' qualification for doing the modelling is that they did it in the Murray-Darling basin! They just don't know anything about the particularities of the Indus River. (A Sindhi Federal Official, 11/09/2015)

The national-scale experts look towards the international scale and draw inspiration from their selective understanding of the experiences in other basins, for example, the Murray-Darling. A national scale aligned with the global developmental scale is the aspiration of the dam builders. The Sindhi official is pointing to the sobering basin and local-scale physical reality of the river which renders moot the comparison with the Murray-Darling basin.

It is not just Sindh that is vehemently opposed to KBD. The provinces of KP and Balochistan are also opposed to it, especially the ethno-nationalist elements in the two provinces. In KP the opposition to KBD is based upon fears of the submergence of the province's rich farmland in Mardan, Nowshehra and Swabi districts and the possible saline water intrusion into the groundwater of the agriculturally rich Pashtun lands. Balochistan's objection is out of solidarity with the Sindhi nationalists as the following articulation illustrates:

With regard to Kalabagh dam, Balochistan has such a small share of the water, that it is not really an issue for us. We go along with the Sindhi nationalists out of ideological solidarity. Nobody can refuse that not building newer dams will cause a water crisis. Right now the lack of trust is the issue, and we have not been able to build the trust. Sindhis believe that if you build the dam you will take away their water. Who will build the trust and which political party will do it, is the question. The major issue is confidence building between Sindh and Punjab. (Hasil Bizenjo, President National Party, 04/08/15)

To sum up, then, the core issue in inter-provincial water conflict is of scalar disconnect between national and provincial visions of benefits and hazards. But beyond that are also competing visions of what Pakistani polity is and should be. On the one hand is a religious belief in national scales of modernist development, where Punjab's interest is Pakistan's interest, while the Sindh's interest is just that – Sindh's interest – or KP's interest is its parochial interest.

I tweeted once that the 3 smaller provinces should get their fair share of the economic corridor. I was put down by people when everybody said 'Mr. Palijo you should think of Pakistan first and not of provinces.' That is the attitude that I will face in the supreme court [if I were to try to litigate Sindh's water grievances against Punjab]. (Ayaz Latif Paleejo, 23/04/2015)

The point of this section is not so much to pronounce judgement on the economic, engineering or hydrological case for or against KBD, but rather to capture the contours of the debate on it. And for me, those contours of the debate make for a depressing reading. To get emotional about an infrastructure project like a dam is like getting emotional about an engine fan belt. They are both means to an end, but in Pakistan KBD has become an end in itself.

Conclusion: Mitigating inter-provincial water conflict

The inter-provincial water conflict is enacted at the expert level, between provincial- and national-scale experts and then again by activists at the national and provincial/regional scales. The scalar production of national and provincial hazardscapes is integral to sub-national scale water conflict. Normal/Average water flows define expert developmental visions and threat perceptions. Average is a predictive model of the data and that model has come to be hegemonic in defining the reality of water management and conflict in the Indus Basin as it has in many others. Expert developmental visions seek maximum returns at the national scale under a stable/normal/average water availability conditions. The expert gaze also perceives hazards insofar as they are deviations from the modelled normal water availability conditions, especially at the national scale. The expert technocratic view at the provincial scale is no different in its allegiance to the average predictive model. Within the boundaries of that model, then, the conflict becomes about experts using and abusing the model to gain advantage for their provinces. The questions of substantive equity, ecology and identity take a backseat in these tactical manoeuvres between provincial water bureaucracies.

The overtly political and politicized hydropolitics links water conflict to wider contestations over identity, ecology and place making. The provincial scale of Sindh is produced through a hazardous view of Punjab's intentions towards Indus River systems' waters. The provincial scale of water in the Punjab is equated with the national scale of Pakistan and then produced through articulations of national water scarcity risk and developmental visions of neoliberal and industrial development exemplified by the West and Pakistan's immediate neighbours of China and India. Within these contestations large dams, especially the KBD, come to occupy a pivotal position. They are integral to producing a national-scale hydro-hazardscape characterized by the risk of India's water diversions and water scarcity from climate change. In Sindh, KBD is equally a perceived threat at the sub-national scale that subverts Sindh's self-production as an autonomous cultural, political and ecological entity.

The most likely scenario in the case of inter-provincial conflicts is for present trends to continue. The present is defined by hegemonic models for viewing water that not only perpetuate conflict along narrow technocratic lines but also preclude water conflict management in a wider sense. Water is imbricated with the questions of identity, ecology and place making. The average/normal model of viewing water produces a blind spot that won't allow the expert mind to listen to or see the activist or subaltern subjectivity regarding water.

No matter how much the water experts insist and even demand that water conflict be treated in a technocratic register, it needs to be dealt with in a political register (Aijaz & Akhter, 2020). Indeed, water conflict, more than many other factors, is proving more corrosive to the federation of Pakistan, as this chapter may illustrate. Inter-provincial water conflict is fundamentally about competing visions of what type of polity and state Pakistan should be. The tensions between the analytical and experiential, and disconnects between the normal and hazardous, views of human–environment relations propel water conflict. Productions of the international transboundary or sub-national scale water conflict are but two exemplars of these tensions and disconnects – besides others.

3

Local-scale water conflict over surface and groundwater in rural Pakistan

Local-scale conflict has not received as much attention in water conflict literature than the better known transboundary and to a much lesser extent the sub-national inter-provincial-scale water conflict. The persistent question at the local scale, too, continues to be of whether it is conflict 'over' or 'through' water. Given water's fluid materiality, its persistence in linking and traversing multiple scales, it is useful to consider how local water conflict are outcomes of cross-scalar processes and are a feature of unique hazardscapes. I will focus in this chapter on conflict around canal water supplies in the Punjab and the groundwater conflict conveyed through the ancient Karez system (underground aqueducts) and tubewells in the highlands of Balochistan, Pakistan. The water conflicts at the local scale in the irrigated plains are a structural outcome not only of the history of the canal colony development but also of its geographical design. Similarly, the conflict over groundwater in Balochistan is an outcome of contemporary development imaginaries that link local Karez communities with the global developmental enterprise. Famine was the signature hazard afflicting the Punjab plains against which the canal colonies were established. The hazardscape of irrigated agriculture was defined not just by the elite concern with famine prevention but also by its imperatives for imperial social control and demonstration of superior western science to potentially recalcitrant 'natives' (Gilmartin, 1994). For groundwater conflict in Balochistan I will argue that the initial impetus was demand-based water management and productivity maximization. But over a period of time drought management has also come to be the signature hazard defining the hazardscape in which the conflict over groundwater is embedded (Mustafa, 2013). I shall briefly outline the historical antecedents to canal irrigation development in the Punjab and groundwater development in Balochistan, to contextualize the subsequent discussion on each of the two instances of water conflict, and their role in producing unstable scalar category of the local.

The historico-institutional context of canal and Karez irrigation in Pakistan

Most water-related narratives about Pakistan focus on the irrigated plains of Punjab and Sindh, and rightly so – after all, vast majority of the country's population and the agricultural productivity comes from those plains. This chapter, too, appropriately devotes considerable space to the local-level conflict in the canal colonies. But the case of Karez irrigation in the highlands of Balochistan is also offered as a comparative case of local-level water conflict to draw out the specific themes of scalar production and cross-scalar linkages within hydro-hazardscapes. I have covered the historical and institutional context of Karez irrigation in considerable detail elsewhere (e.g. see Mustafa, 2013, 2007); therefore, I shall very briefly revisit the basic details. Karez is an ancient system that involves an underground water channel tapping the groundwater passively and then conveying the water by gravity to the downhill water users living past the daylight point (Figure 3.1). The system is also called Qanat in Syria, Falaj in the Arabian Peninsula and Foggara in North Africa. It exists across southwest Asia, central Asia, North Africa and as far afield as Mexico. The system typically involves transferrable and fungible time-denominated water rights. The Karez system in Balochistan, for example, typically has twelve to thirty-six-day cycle of water rights where water users may own rights ranging from a few minutes to entire twenty-four hours to multiple day cycles. One of the equitable features of the Karez system is that everyone who owns the first parcel of land after the daylight point also owns the last parcel of land. Whoever owns the second parcel of land also owns the second to last parcel of land along the water course after the daylight point and so on. This particular feature provides equal incentive to every water user to contribute to the maintenance of the entire water course instead of the typical conflict between the head and the tail-end water user situation so prevalent in the surface irrigation water systems, as I shall discuss later.

The individual Karezes in Balochistan range from a few decades to hundreds of years old. The water distribution system is purely customary. Every Karez has a water manager called Mir-e-aab who maintains record of water rights, arbitrates any water conflict and organizes maintenance for the infrastructure. This purely indigenous water system is now increasingly under threat from the state-introduced high-powered water pumps called tubewells which are lowering the water table, sometimes below the mother well level, thereby leaving many Karezes dry and hence their shareholder water users

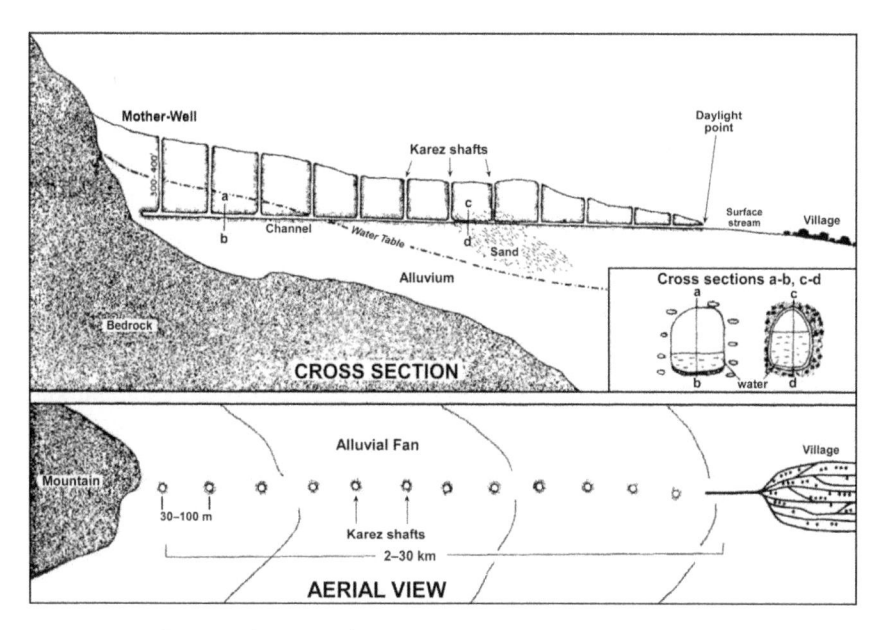

Figure 3.1 Schematic diagram of a Karez.

out of their water right. I shall get to the dynamics of this conflict in the next section.

The famous Pakistani surface irrigation system constituted a few inundation canals built by the Mughal Empire in the seventeen century. Subsequently, it was the British Empire that built the perennial irrigation canal system that was subsequently extended into its present areal extent of 44 million hectares at present (Michel, 1967). The very motivation and the subsequent design of the system were to solidify imperial control over the recently acquired province of Punjab in the 1860s. The idea was to settle the so-called loyal and industrious castes in the colonial mythology at the heads of water course, and to create a feudal elite which would owe its allegiance to the British Empire and hence be the instrument of control over the rest of the population (Ali, 1988).

The physical system is gravity based, where water diverted from the mainstem rivers gets diverted to canals, from which radiate out distributary and then often minor canals. The system has low conveyance efficiency where only 30 per cent of the water diverted from the main stem rivers gets to the root zone of the crops. The rest of the water is either lost to evaporation or to seepage, which is not a problem in the fresh groundwater zone, because it is pumped right back out for demand-based water management, but in saline groundwater zones, that becomes a problem. Tubewells have become increasingly important in this surface irrigation system where, just in Punjab province, there are more than a million privately owned tubewells sometimes supplying close to 80 per cent of the crop water requirements (Shah, 2007). The tubewells are particularly important because of the fact that the original system was designed for 64 per cent cropping intensities, where scarcity was to be shared equally between all water users. Today the cropping intensities are up to 150–200 per cent.[1] Even then, as now, the scarcity was not equally shared because water courses at the head of canals and water users at the heads of water courses are at a distinct advantage because of conveyance losses from seepage and evaporation. But the water charges, although static since 1970, are the same between the head and tails of water courses. An hour of water right volumetrically means different things between the head and tail ends of water courses, and canals because of conveyance losses. This is a feature to be kept in mind as the narrative progresses (Mustafa, 2013).

Village water courses almost always emerge from the distributary and minor canals and almost never from the main canals. The village water course inlet is called a *Moga*, and they are of specific sizes allowing for specific flows dependent upon the acreage that they are supposed to irrigate below the moga. The system is administered by the provincial irrigation departments which are mostly manned by civil engineers and tend to have a technocratic approach towards managing the system, except when they systematically do favours for the larger landowners and water users (Mustafa, 2013). The water distribution is based upon a fixed-time rotational system, where every farmer along a water course has a fixed time right to the entire flow of the water course proportionate to the amount of land that the person holds. Once a farmer along the water course is finished diverting water for the designated amount of time, the next farmer along the water course may divert the entire flow proportionate to his or her amount of land to be irrigated. In this respect the water distribution system is not that different from the Karez system, but as I will illustrate in the following sections, the local-scale conflicts are illustrative of the historical constructions of the local-scale water use and how that 'local scale' is being dragooned into national-scale imperatives of increased productivity as well as political fault lines.

---

[1] Assuming two cropping seasons, if 100 per cent of the land is cultivated twice a year, then the cropping intensity would be 200 per cent. The 64 per cent cropping intensity means that the farmer was expected to cultivate about a third of the land twice a year.

Pulling the local into global: Groundwater conflict in Balochistan

Irrigated agriculture and the arid highlands of Balochistan, Southern Afghanistan and Iran had always depended upon groundwater. The conduit for conveyance of groundwater to the surface had been the Karezes for millennia. The Karez system was a quintessential local-level infrastructure mostly locally financed and managed, and supporting local livelihood and social systems. Very occasionally local tribal rulers would patronize Karez construction and/or maintenance as a way of materially claiming a stake in an essential piece of infrastructure and thereby legitimizing suzerainty over vassal territories. Karezes had always been more than just a water infrastructure. They were the locus of social relations, livelihood and cultural continuity for the communities dependent upon them, for example,

> Karezes were a great source of social and communal life for us village folks. People would sit on their sides and discuss their issues and find solutions to their problems. (Karez shareholder, Kunghar Karez, July 2004)
>
> Our elders tell us that in olden times there was so much water in the karez that people working far away in the fields would receive their mid-day meals through the karez. The women folk would put food in a buoyant sealed container called 'Danko', and put it in the karez near them. The karez would then take the containers to the farmers working downstream. Now those were the days! (Karez shareholder2, Kunghar Karez, Balochistan, June 2004)

Karezes require considerable maintenance annually for them to remain operational. Mobilizing the labour for that maintenance requires strong social capital and organizational skills, particularly to keep the process equitable and fair to all the community members. It was the social and cultural capital which gelled around the operations and maintenance of Karezes, which contributed towards community coherence in the arid and semi-arid highlands of Balochistan. But things changed relatively quickly as the developmental state made inroads into previously isolated communities at the geographical periphery of the Pakistani state. Starting in the late 1970s, international developmental actors such as the USAID arrived in Balochistan and immediately deemed the perpetually flowing Karez system to be antiquated and wasteful, without inquiring into the important social and cultural functions that the infrastructure served, not to mention the important equitable features of the system. The 'antiquated' system was to be replaced by demand-based, state-subsidized tubewells operated by state-subsidized flat-rate electricity.

One of the features of the Karez system is an almost universal *harim* rule whereby no other mother well, pumping or potentially polluting activity can be undertaken within a specified distance (typically 1,200 feet) of the mother well of a Karez. As some communities through political patronage started taking advantage of the state-subsidized tubewells, the groundwater started declining.

> There is never conflict on Karez water. There is conflict on communal land use and tubewell installation near the Karez, e.g., the other tribe's tubewell is close to our Karez and therefore there is conflict. (Water user, Soghai Allahdadzai, Balochistan)

The drought between 1997 and 2003 led to accelerated installation of tubewells as a drought management technology. In the Baloch hazardscape the modern technology of tubewells was deemed to be the solution to water scarcity and drought by the government and international technocrats. This technical solution was to have disastrous consequences for the Karez system, as the water table was lowered below the level of the mother well, with many Karezes going dry. Even when the Karezes were often replaced with tubewell irrigation, new water-related conflicts were to emerge as a result of this transition, for example,

> Tubewells benefit individuals, karez benefits society and binds community together. Karez was also a mechanism for conflict resolution, though there is no future for it. Contradictory government policies and bias towards tubewells is to blame for it. . . . This is by the way an atypical perception [in my department]. You see we in the government are into tangibles, e.g., how much money spent and tubewells installed. (An Assistant Director, GoB Agriculture Department)

> Wherever a tubewell is installed next to a karez the karez shareholders lose out. The tubewell owner wins out with higher profits. But a tubewell is owned by an individual from which, two or three people are earning their living, but a karez is communally owned from which 500–1000 people may be earning their living. So, you figure out that when a tubewell gives an individual benefit, how many loose out. (Water user, Bangi Karez)

> Large landowning farmers have benefited from the tubewell and have provided work for poor people. We [small farmers] are the ones who have suffered because our karez water is reduced, our lands are lying fallow, and unemployment has increased beyond all limits. (Woman water user, Bangi Karez)

The remarks here reflect a classic case of individualizing the profit and replacing the communal ethos which are structurally integral to Karez irrigation with the notion of enhanced productivity for private profit (Mustafa & Qazi, 2007). This is also an instance of the local-scale productive activity being pulled into national- and global-scale capitalist markets, where productivity and profit are at a premium instead of local-scale values of community and social coherence. The conflict in this instance is not only between the Karez shareholders and tubewell owners but also at times within communities where after the drying of the Karez some communities physically installed a tubewell within the Karez and superimposed the Karez water rights onto the tubewell water, with similar rotational schedules as in the Karez. In such instances water conflict became much more acute than had ever been the case with the Karez water, for example,

> Sometimes there is conflict on tubewell water, e.g., if today is our turn to irrigate from the tubewell, but we do not have the money for the diesel or the tubewell is not working, we cannot irrigate. The following day when it is another member's turn to irrigate, we would argue that since we did not get water on our turn, we should get the water today. And this becomes the cause for conflict. (woman water user, Yakub Karez)

As the quote illustrates the rotational irrigation schedule that worked with the perpetually flowing Karez became a source of conflict between the shareholders of tubewell water, as electricity failure, lack of fuel in case of diesel-powered pumps and breakdowns meant that people who weren't able to get their designated turn would either have to forfeit their watering right in that cycle or get into conflict with the person who had the next turn. The consequence was greater strain on community conflict management mechanisms to resolve such conflicts (Mustafa, 2007).

The tubewell technology with its dependence upon extra local energy sources, for example, electricity and diesel fuel, technical expertise for repair and maintenance, and the production systems which favour larger-scale commercial agricultural production, is instrumental in undermining the local-scale livelihoods and social structures. Drought and uncertain precipitation are by definition endemic to arid and semi-arid regions in which the Karez system operates. Yet the system sustained human habitation through millennia and withered many a droughts in the past. Yet at the end of the twentieth century, it was deemed to be too antiquated and wasteful, and the tubewell was instead promoted as a panacea for lower productivity and drought. The consequence was indeed enhanced productivity, largely to the advantage of the large landowners and the disadvantage of small farmers. The erosion of the local scale was complete, and the production of a national- and international-scale agricultural economy perfected as a consequence. Through migration and diversification of livelihoods into the service sector, the populace was able to reinsert itself into a new economy more integrated into the national scale (van Steenbergen et al., 2015). Comparable processes are also at play in surface irrigation conflict in the canal colonies of the Punjab, as I discuss in the following section.

Producing the local from national: Surface irrigation water conflict

The key piece of legislation governing water management in the irrigated plains of Pakistan is the Canal and Drainage Act (1873). The act defines two most commonly prosecuted offences to be under Section 68, which typically speaks to taking water out of turn (*Warashikni*) and hence causing conflict between water users, and Section 33, which speaks to tampering with the moga, or taking unauthorized water directly from the canal (Mustafa, 2001). At the time of independence more than 90 per cent of the village water courses had *kacchi warabandi* (locally defined water distribution schedules). Contemporaneously more than 95 per cent of the village water courses, however, have *pukki warabandi* (irrigation department defined and recognized official water distribution schedule). The transition has been from the local-scale customary water rights to the provincial scale sanctioned and officially authorized water rights in the shape of *pukki warabandi*. But as I have documented elsewhere, for example, Mustafa (2001), much of the prosecutorial zest of the authorities for *warashikni* (taking water out of turn) is directed heavily towards small-farmer-dominated canal commands. Not because large farmers are necessarily more honest, but because the irrigation authorities cannot take on the large farmers who often constitute the political leadership of the area as well. The provincial scale of water management has to defer to local-scale contours of social power in managing water conflict. I documented almost identical trends in the moga-tampering-related offences.

Water conflict at the local scale is, however, often embedded in local- to national-scale caste, ethnicity and political-affiliation-related fault lines. For example, as documented by Milan Karner in Mustafa et al. (2017) water conflicts are rarely about water. Typically, water becomes a medium through which personal, familial and even political conflicts are played out and negotiated, for example, as per an irrigation department official quoted in Mustafa et al. (2017):

> In 100 percent of the [water-related] cases that I have prosecuted, the grievances is something else, but they make cutting off the water as a way of settling disputes.

> In my 30-years-experience, there has never been an instant of water being the sole cause of conflict. It has always been the symptom, and a weapon for settling other disputes. (SE PID, 27/04/2015)

Milan Karner in Mustafa et al. (2017: 35) reports such an instance of water being used as a stand in for larger conflicts over political affiliations as follows:

> a water dispute about the reallocation of left-over surplus irrigation water (*nikaal*) among farmers at the tail-end of a village watercourse in Sahiwal District broke out after the recipient of the *nikaal* stopped the practice of sharing the water with two neighbouring families. This took place following the neighbouring families' change of political party affiliations. While they held no rights to the water, the families felt entitled to a continued share of it in view of the long period that this practice had been in place. A first plea of the original recipients' decision to reallocate the surplus water was rejected by an irrigation official several years ago, but a second one recently yielded a favourable decision that effectively reallocated the *nikaal* to one of the other families and hence away from the original recipient of the water. It is not known why the official settled the water dispute simply by reallocating the entirety of the surplus water to one single family, hence merely recreating the previous situation of unequal surplus water allocation – that gave rise to the conflict in the first place – with another beneficiary. It is, however, indicative of how deeply water access and politics are entangled in rural agricultural communities in Pakistan and how personal and political influence undermine institutional and judicial processes. It is noteworthy that while a change of party affiliations may have triggered the conflict in the first place, it may have also led to the eventual resolution of the conflict in the contesting family's favor. The president of the farmers' organization in charge of the mediation happened to have the same party affiliation as the people who had been deprived of the water by their neighbour.

Karner in Mustafa et al.(2017) further reports an instance of the murder of a village headman in Sahiwal district in central Punjab, supposedly over taking water out of turn. But upon further inquiries he learnt that the murder was also driven by deeper familial conflict about switching of political allegiances by one of the parties in the conflict. Although, Milan Karner rightly notes (p. 35) that it is impossible to tell whether the water or the political conflict was more of a factor in the murder, but this example and the one about *nikaal* do illustrate two things. In the first instance they illustrate how water and politics are invested with issues of pride and honour in the Pakistan rural society. Second, they illustrate how national- and provincial-scale political conflict ends up defining the hydro-hazardscapes of water scarcity at the local scale. It is an instance of the provincial- and national-scale conflicts being violently reproduced through water at the local scale.

Historically the moga tampering and *warashikni* cases had always been prosecuted by the Sub-Divisional Canal Officers (SDOs) and Executive Engineers (Xens), who also have magisterial powers. For the past two decades, though, many canal commands have been turned over to water user associations (WUAs) under pressure from the international donors, who made it a condition of their continued lending and support in the water sector of Pakistan. In this instance the production of the local scale of water management in opposition to the provincial bureaucratic scale of water management is, in fact, towards delegating power to local-level farmer organizations and thereby consolidating a local scale of water management. But even then there is anecdotal evidence based upon conversations with irrigation staff, as well as personal observations in southern Punjab and northern Sindh that most canal commands that were turned over to WUAs were typically dominated by large farmers. For example, Koranga Distributary in the Sidhnai Barrage area in southern Punjab, which is dominated by some of the largest and politically powerful farmers in the region, was turned over to WUAs and not the neighbouring Darkhana Distributary, which has a small-farmer-dominated water user profile (Mustafa, 2013). The intent of the WUA reforms was to move away from the bureaucratic model and its accompanied corrupt practices, and instead hand over control to farmers who could arbitrate local conflict and collectively negotiate water supply with the irrigation department. Instead, from the very inception of the WUAs in the late 1990s (Bandaragoda, 1999), it was noted that large farmers tend to dominate such associations and generally end up replicating their power to influence the bureaucracy to their advantage, by directly dominating the WUAs. The efficiency and equity scenarios envisaged at the international scale through WUAs do not necessarily translate into equity at the local scale, as has been documented elsewhere as well, for example, see Mustafa et al. (2016) for WUAs in the Jordan valley. The issue of elite capture is a live one, where power and influence of the large farmers force many small farmers to stay quiet in the face of water theft by others, for example, as documented by Milan Karner in Mustafa et al. (2017: 33):

> I'm a poor farmer. I approach the chairman of the *khal panchayat* (the water course committee) [in cases of water theft]. Some issues are resolved, but not all. If the value of the stolen water is Rs. 500, the bribe to the police would still be Rs. 1000, so I'd rather keep silent. I would also have to buy the *lumberdar* dinner to get him to act, which would easily cost me another Rs 500. (Farmer in Sahiwal Division, 24/07/15)

Between the provincial-scale water bureaucracies which are often obsessed with technocratic modes of water management and typically go to great lengths to isolate themselves from the civil society and especially small farmers and water users (Mustafa, 2002) and the large farmers that dominate WUAs, the smaller water users tend to be resigned to silently suffer the consequences of the injustices they suffer. Small farmers typically engage with conflict only with other small farmers and almost never with the large farmers, let alone the bureaucracy.

Local-scale water conflict is almost structurally integral to the Pakistani irrigation system. As noted earlier the irrigation system was designed for 64 per cent cropping intensities while the present-day cropping intensities stand at 150–200 per cent, thanks largely to chemical fertilizers and the demands of a monetized green revolution agricultural economy. The farmers at the heads of canals routinely tamper with mogas to get more than their fair share of water at the expense of downstream water users. The farmers at heads of water courses steal water at the expense of farmers in their own communities as well, provided they are smaller or of equal size to them in power. The diesel pumps and electric tubewell technology has allowed tail-end farmers to compensate for the water lost with groundwater. But there, too, larger farmers can bear the capital expense of installing deep electric tubewells, as well as get around the high operating cost, by simply stealing electricity. The smaller farmers who cannot afford the capital expense have to content themselves with low-powered, portable diesel-operated pumps. With increasing groundwater draw down, that race to the bottom is again negatively affecting the smaller farmers at the tail ends of water courses in the first instance. Little wonder then that farm abandonment by smaller farmers and rural to urban migration is the highest for Pakistan in Asia (Mustafa & Sawas, 2013).

Conclusion and future pathways

Irrigation water conflicts at the micro-scale will remain an issue of considerable concern in Pakistan in view of the disproportionate share of the country's water resources allotted to irrigation and the continuing importance of irrigated agriculture for the national economy. Other conflict constellations at the sub-national level may eclipse local irrigation water conflicts in terms of scope and the attention that they receive from both politicians and the public, but the unequal access, allocation and availability of irrigation water and attendant conflicts will continue to permeate rural life and immediately affect the livelihoods of a sizeable share of Pakistan's population. The institutional factors that are conducive to irrigation water conflicts, such as the lack of accountability on the part of irrigation officials and the impunity of powerful landlords, cannot be viewed – nor transformed – separately from the wider bureaucratic and political culture of Pakistan. The performance of participatory water management institutions such as farmers' organizations is indicative of the constraints that the wider structural context imposes on individual initiatives or attempts at fixing parts of the current system.

The local-scale hydro-hazardscapes feature the overwhelming power imbalance between the international donors, the irrigation department officials, the large farmers on the one hand and the smaller water users on the other. It is not a coincidence that the highest electricity bill default rates are in regions with highest agricultural dependency upon groundwater. The electric supply company in Balochistan Quesco, for example, has the highest default rate in Pakistan (Naveed, 2019). In case of the decline of the Karez system the international-scale donor pressure and national developmental visions undermined the local scale of Karez irrigation, and replaced it with the mostly regional- and national-scale dependency on the tubewell technology. The technology further re-programmes the rural economy to produce for a national- and international-scale agricultural market.

In case of the surface irrigation system in the canal colonies, the provincial-scale water bureaucracies intervened, mostly to further reinforce the local-scale power of the large landowners. The system that worked on corruption and patronage, thanks to the international donors like the World Bank, has now been formalized through WUAs in the large-farmer-dominated canal commands. Even the fig leaf of the provincial government standing as an honest broker in water conflict is removed in this instance.

The scalar politics do not have to insert themselves perversely into the water conflict at the local scale. The provincial and national scale has to be contested and then recast into a different mould from its unidirectional obsession with reinventing the local as a contributor to the technologically advanced mass production, mass consumption society. That larger recasting of the national or even international will take time, but doesn't mean that it cannot be worked towards. As an interim policy intervention there is potential for mitigating conflict over surface water by rationalizing the water charges by location along the water course instead of the present flat rates. Further, participatory water reforms are good in principle, but they are unlikely to realize their potential for equity in large-farmer-dominated canal commands. Their equity potential is precisely to be realized in small-farmer-dominated canal commands. In addition, leveraging the social capital built around Karezes to negotiate water conflict could pay rich dividends. In addition, it must be realized that the long-term sustainability of irrigated agriculture is not possible through tubewell-based overdraft of water, which is what's happening in Punjab and Balochistan.

In the short term, the implementation of on-farm water management methods and water conservation technologies that is being promoted by the government and non-governmental organizations alike (such as the laser-levelling of fields and the introduction of drip irrigation) may allow the continuation of agricultural production. In the long run, however, Pakistan will not be able to avoid addressing the larger structural and institutional issues that plague its agricultural economy, particularly the gap between the designed and the actual cropping intensities of the irrigation system and the cultivation of water-intensive cash crops in water-stressed areas. These pose excessive demands on the available water supplies and are bound to exacerbate conflicts within and between farming communities as agricultural production is increasing, and the available quantities of water for irrigation remain constant (or even decrease). The fact that irrigation water conflicts in rural Punjab or Balochistan cannot be detached from other contestations over water and questions of justice at the sub-national level becomes evident in the words of an executive engineer at Sukkur Barrage in Sindh: "They lament Kalabagh dam on the Indus; there is a Kalabagh Dam on every watercourse in Sindh." It would be well if the debate about sub-national water conflicts in Pakistan reflected this, and irrigation water conflicts at the micro-scale were given the careful attention they deserve.

4

Contested hazards in local hazardscapes
From floods to pollution

Floods are blind and cannot see. But along the way society and history have conspired to become their eyes. The lived geography of floods is consequently characterized by deep fissures along ethnicity, class, age and gender lines (Mustafa & Gioli, 2015). Less dramatic and photogenic than floods, water pollution too differentially impacts the most disadvantaged, as we will demonstrate in this chapter. Like floods, water pollution too is steeped in deeper conflicts over visions of development and whose life spaces are valued by the society. Conflict over and through water is not just about quantities as is popularly understood but also about how its associated hazards are distributed. In a highly regulated hydrological system like Pakistan's, there is nothing natural left about the timing, location and intensity of the floods in the basin. The human influence on water pollution is of course obvious, being that it is a deeply modern phenomenon inextricably linked with the development of irrigated agriculture in the nineteenth and the twentieth centuries described in the previous chapters. In this joint treatment of floods and water pollution with specific reference to the case of Manchar Lake in Sindh, I hope to highlight how the fractal production of scales manifests itself even in the seemingly local hazards like floods and pollution. Many in the Pakistani media may have discovered the politically contested nature of decisions to flood one or another region along the rivers during the massive 2010 floods, but the political influence on irrigation and flood management is as old as the system itself. Similarly, the pollution and its profound effects on the lives and livelihoods of fisher communities in the Indus are also an outcome of a historical process that privileged agricultural livelihoods over other livelihoods systems, especially fisheries in the basin.

After describing the physical and historical trends in flood and drainage management in Pakistan, the chapter will turn to the scalar politics of flood hazard in Pakistan. It will then contrast flood conflicts within the upstream

Punjab and the downstream Sindh province to segue into a discussion of drainage and accompanying pollution in the context of freshwater fisheries in Sindh. The discussion on drainage and water pollution will be with specific reference again to scalar politics of drainage as well as conflict between agricultural and fisher interests in the Indus, as an example of contestations inherent in the Indus hazardscape.

The historico-physical geography of floods in Pakistan

It has been mentioned before that the Indus River has some of the highest silt loads in the world. It is the deposition of that silt that has formed one of the most fertile alluvial plains on the planet, enabling the establishment of one of the oldest civilizations in the world. The flood cycle was, and still is, integral to the maintenance of the fertility of the alluvial plains, and our discussion here is largely limited to the riverine flooding to the exclusion of local-level hill torrents and flash floods in the mountainous West and dry lands of the country. Rivers are dynamic systems, and in a sense ideal hazardscapes, with uncertainty integral to their cycles of deposition, erosion and channel changes. That dynamism of the rivers on the one hand provides the bounty of water and fertilizing alluvium for the soil. But on the other hand, that dynamism can be quite inconvenient if one wants to put an irrigation system on it, complete with weirs and barrages and canal off-takes. One cannot build a barrage at a particular point in the river, only to have the river change its course and move away from that location. Furthermore, if one moves from a flexible communal property regime towards a static individual landed property regime, which is, in fact, what happened with land settlement in the basin in the nineteenth century, it becomes doubly imperative to protect individual properties from getting submerged or eroded by moving rivers. Consequently, the Indus Basin Rivers had to be controlled and hemmed in between levees to make sure that they continued to flow within channels that were convenient for their human managers, but not necessarily for their natural regimes.

In the gently sloping plains of the Punjab and Sindh, primordial rivers would spread out their floods over large areas and most of the time the human societies would not only be adjusted to those cycles but in fact be dependent upon them for their livelihoods. However, as the nineteenth-century canal colonies were established, the very blessing of floods turned into a hazard. As the rivers were constrained to flow within specific channels, and as more and more water was diverted away from them, in the plains at least, the rivers lost their capacity to transport their sediment load and spread it over wide areas, and in fact started depositing them within their river beds – thereby markedly reducing channel capacity (see Figure 4.1). This development is what Mustafa and Wrathall (2011) call a Faustian Bargain, whereby the human societies exchanged the dynamism of the river with its frequent low- to moderate-intensity floods for higher productivity, static agricultural holdings and low-frequency but high-intensity floods. This broad paradigm of flood control through river engineering is concretized through the operational imperatives of protecting the barrages and dams on the rivers, each one of which has a safe design capacity, often based upon relatively limited historical data at the time of their construction, compared to the data that is available since the original construction of the infrastructure (Mustafa & Wescoat, 1997; Mustafa & Wrathall, 2011). Historically as well as contemporaneously, every time the safe design capacity of the barrages is exceeded, the upstream levees are deliberately breached to relieve the pressure on the barrages, at least in the Punjab. Where those levees will be breached has generally been decided at the time of the construction of the barrages, and the inundation zones are selected – and well known – in advance. This is one important humanly modified physical aspect of flood management in the Indus Basin that must be borne in mind, as the narrative progresses.

Almost invariably, the levee breaches upstream from dams are on the right bank of the river. In fact, there are twenty-three such designated breaching and

Figure 4.1 The bed of the Ravi River downstream of the Sidhnai Barrage with extensive silt deposition within the channel.

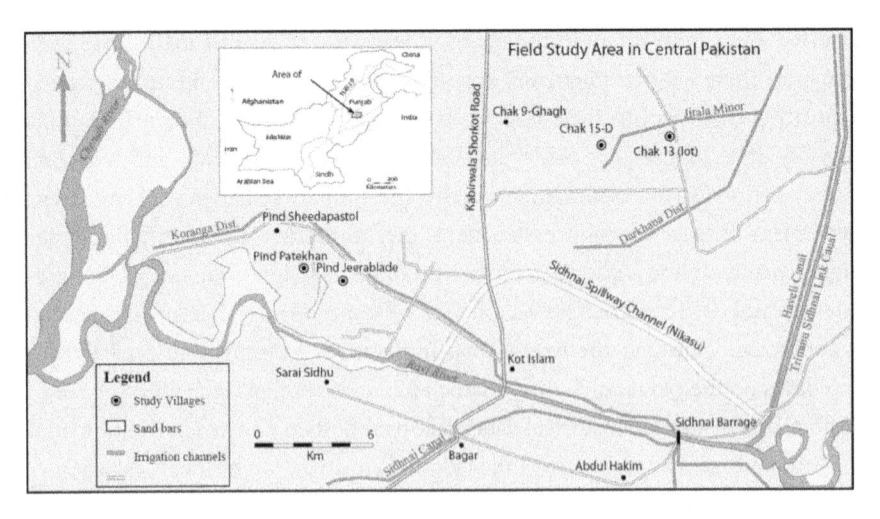

Figure 4.2 A map of the inundation zone on the right bank of the Ravi River.

inundation sites in Pakistan. Figure 4.2 is a map of one such inundation site upstream from the Sidhnai Barrage on the Ravi River. The assumption behind keeping all the inundation zones on the right back of the rivers is based upon the Coriolis effect,[1] that is, the idea that on the right bank the water will be deflected back into the main channel, whereas on the left bank water will keep flowing indefinitely away from the channels. Although the assumption is theoretically correct, the inundation zones also have levees, roads and road berms, railway lines and canals cross-cutting through them and serving as barriers to where the water wants to flow (see Figure 4.3). The consequence is that the water, once it has left the main channel, gets stuck between all the human-made barriers and sometimes does not recede for months, with the local population camped on the roads and high grounds suffering from water-borne diseases *ad interim*. It is for this reason I argue that rather than inundation, it is the drainage of flood waters that is the main physical challenge in the Indus Basin. Good drainage can ensure that the quiet suffering of the flood victims, long after the TV cameras are gone in the immediate aftermath of the inundation, is mitigated.

The issue of flood drainage becomes particularly acute in Sindh. Sindh's good fortune thus far has been that the headwaters of the main stem Indus River do not fall under the summer monsoon influence (Yu et al., 2013). It should be noted that there are two flood seasons in Pakistan: a minor one following the snowmelt on the upper reaches of the rivers in April and May and the major one from July

---

[1] The Coriolis effect is the apparent deflection of every moving thing to the left in the Northern Hemisphere and to the right in the Southern Hemisphere as a result of the rotation of the Earth.

Figure 4.3 The 2010 inundation situation in Muzaffargarh district clearly showing the road acting as a barrier to the drainage of the flood peak.

through September caused by monsoon precipitation. Since the eastern tributaries of the Indus are mainly affected by monsoon floods, historically, by the time those flood peaks entered the main stem Indus River downstream of Panjnad, they were easily absorbed by the much larger river channel of the Indus. However, that good fortune may not last into a climate change future. The 2010 flood was a result of unusual monsoon rains in the Kabul and Indus River valleys (Mustafa & Wrathall, 2011). Before 2010, a major flood in the Indus mainstem was a result of glacial lake outburst floods (GLOF) back in 1929. With an increasing probability of downstream Sindh being affected by the floods of the Indus, it should be noted that almost the entire length of the Indus River in Sindh is located within levees. It is also in Sindh that the question of drainage becomes more urgent beyond just flood water drainage, as we discuss in the following section.

Drainage as a hazard

Sindh province has exceptionally gentle slopes and a number of depressions that have turned into lakes from Indus inundation events. These depressions can be a source of storage of flood water in the Indus, for example, the Manchar Lake which I will be discussing later in this chapter. With the development of the irrigation system, however, the obstructions in the way of the drainage of flood waters have been made worse in Sindh than in Punjab. The lakes that used to be sources of freshwater recharge, fisheries and livelihoods for thousands of people have now been transformed into receptacles for holding saline drainage water, which is channelled to them by internationally funded drainage projects in their vicinities.

It was recognized earlier on that conveying surface water to the agricultural land was one part of the problem. The water once there, also had to go somewhere, if it did not evaporate or seep into the groundwater. In Sindh more than 80 per cent of the groundwater is saline and was found at significant depths before the advent of the surface irrigation system. The seepage water from the surface irrigation raised the saline groundwater almost to surface level, leading to the problem of waterlogged soils. When the groundwater which had reached the surface evaporated under the intense summer sun in South Asia, it would leave behind a layer of salts, conveyed to the surface by the capillary action of the groundwater. The issue of waterlogging and salinity was recognized as a substantial problem earlier on but was not tackled substantively till the 1960s. Pakistan was a close Western ally during the Cold War in the 1950s and 1960s. It was therefore also a recipient of substantial Western financial and technical help during that time under the Indus Basin Development Programme (IDBP) in the water sector, to build replacement infrastructure in the aftermath of the IWT. A highly influential report by the White House Department of Interior Panel on Waterlogging and Salinity in West Pakistan (1964) came to be known as the Revelle Report. The report called for large-scale public tubewells and surface drainage systems to address the waterlogging issue. Despite opposition the report's recommendations came to form the blueprint of how waterlogging issue was to be understood and managed by Pakistani and international water managers (Wescoat et al., 2000).

Local and regional drainage channels were built earlier on during the development of the irrigation system in Sindh. The drainage construction was further accelerated during the IDBP and then subsequently as well under the World Bank-funded Salinity Control and Reclamation Projects (SCARP). Much of the drainage development during SCARPs was based upon provision of tubewells, and then construction of surface drainage channels for carrying away the effluents from agriculture. One such spinal drainage channel that carries effluents from multiple local channels is the Main Nara Valley drain (MNVD), which is part of the larger Right Bank Outfall Drain (RBOD) project started in 1992. Although MNVD was first built in 1932, to carry effluents from rice farming on the right bank of the Indus, it subsequently came to be a spinal drain carrying the effluents from many other regional drainage channels under the RBOD project. The RBOD was originally supposed to dump its effluent load into the Indus River but that was prevented by the environmental lobby in Pakistan.

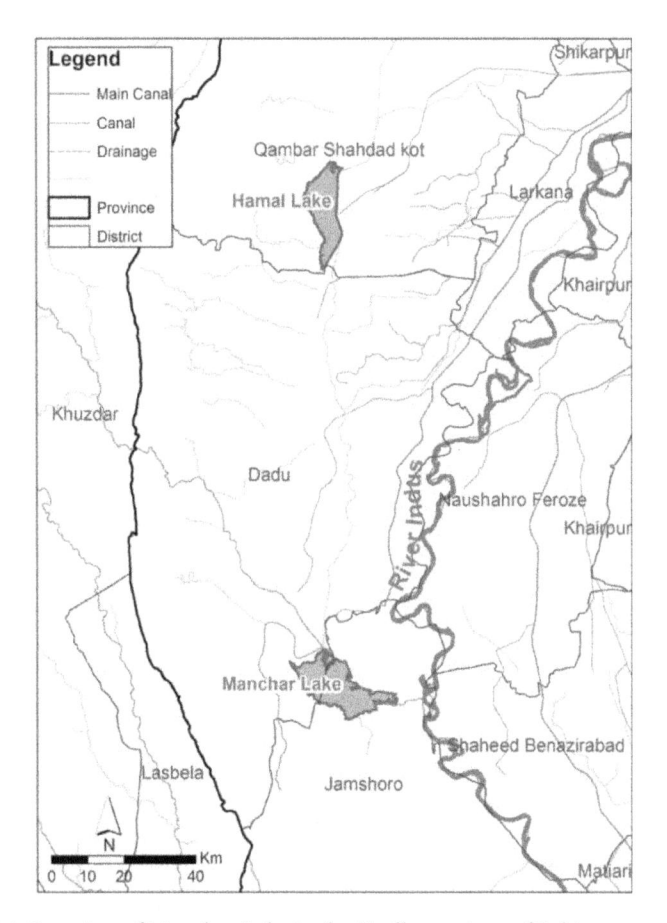

Figure 4.4 Location of Manchar Lake in the Sindh province of Pakistan.

As an interim measure the RBOD project extended the MNVD to Manchar in 1997 and dumped all of the agricultural effluents that it was carrying into the Manchar Lake (Figure 4.4) (Mustafa et al., 2017). The results were catastrophic for the ecology of the Manchar Lake and the people who had come to depend upon that ecology for their livelihoods. This chapter is partially a tale of that catastrophe that has befallen Manchar and conflicts that define the disaster.

The fractal scalar politics in flood and drainage management in Pakistan

At the national scale in Pakistan, as Wescoat et al. (2000) document, the attitude towards flood hazard was primarily one of risk acceptance until the 1973 massive flood event. Since then, a more proactive approach towards flood control has emerged among Pakistan's water managers. The Federal Flood Commission (FFC) was formed in 1977 to provide technical assistance to the provinces in the implementation of flood control projects. The national-scale approach towards flood hazard has largely been centred on engineering interventions and subsequent relief activities by the Pakistani Armed Forces and provincial relief commissioners. To date, there is little understanding of the local- and regional-scale differential vulnerability of the residents of the flood plains to flood hazard on account of class, ethnicity and gender. This theme also gets deeply imbricated in the scalar politics and geopolitics around flood hazard in the country.

In Pakistan domestic floods take on an international geopolitical colour thanks to the rampant rhetoric, mainly espoused by the right wing, in the Urdu-language media, that floods are largely the result of deliberate acts of releasing water during the flood season by the Indian water managers. For example, *Daily Nawa-e-Waqt*, a conservative Urdu newspaper in Pakistan, uses the term *Abee Jarhiyyat* (water aggression) in its editorials to illustrate what it sees as India's deliberate policy of flooding of Pakistan. Others in Pakistan have also used the term 'water bomb' to describe how India manipulates water flows to settle a score with Pakistan. More often than not, papers such as *Nawa-e-Waqt* are repeating this rhetoric based on their generic anti-India posture, depicting the neighbouring country as the enemy using every opportunity to destroy Pakistan.

> India has refused to share with Pakistan the data of water level in rivers, increasing the chances of floods again during this monsoon season, which may lead to deaths of hundreds of people and destruction of infrastructure of billions of rupees. The Indus Commissioner has informed the government that India is refusing to share the data. . . . During the last year, India suddenly discharged water in Chenab and Ravi, which lead to deaths of 367 people in Punjab, damaged 100,000 houses and 700,000 people were displaced. (*Daily Nawa-e-Waqt*, 02/07/2015)

This rhetoric has substantial traction within the Punjab province, where some of the lower-level irrigation officials have expressed the same opinion that India causes floods by deliberately releasing water onto Pakistan. Right wing political parties and militant organizations have also leveraged bilateral water issues with India, particularly controversies about dams and floods, to peddle their own anti-India agenda. A media report quotes Hafiz Saeed, the head of Jamaat-ud-Dawa, as saying:

> India irrigates its deserts and dumps extra water on Pakistan without any warning. . . . If we don't stop India now, Pakistan will continue to face this danger. (Reuters, 17/09/2014)

Even in the more 'moderate' English-language press and the best and the oldest of them, the daily *Dawn*, news reports often hint upon India's malevolence pronounced by a public official as a cause of floods in Pakistan, with the headline, 'Pakistan will ask India to inform before releasing dam water.' The news story states,

> Chairman National Disaster Management Authority (NDMA) Major Gen Muhammad Saeed Aleem told reporters at the conclusion of a two-day national conference held to review monsoon preparedness: 'We have asked Pakistan Commission for Indus Water to take up the matter with India.' He said that in the past the country had suffered huge loss because of the release of dam water by India without any warning. However, the upper state (India) is bound under the Indus Water Treaty to provide information before making such a move. (Khan, 2014)

The above-mentioned statement is factually not correct, because according to the IWT, India is bound to provide water release data to Pakistan as a matter of routine. Now whether that data gets disseminated is an internal Pakistani matter.

The flood hazard as per the above becomes an exercise in enacting an international-scale hydropolitics and in the process solidifying the national scale of flood management in opposition to the alleged Indian misdemeanours in causing floods in the Pakistani Punjab. The exercise is not that different from the national-scale making in the case of water management and distribution, as documented in Chapter 2. The national scale of flood management suggests large-scale infrastructure projects, such as the Kalabagh and other dams for water storage and flood prevention, and a complete neglect of geographies of differential vulnerability to the hazard and local-level spatiality of who gets inundated, and why.

Although there is little evidence to suggest that something comparable to Pakistan–India blame game on flood hazard is operative between Sindh and Punjab, we did find that there was considerable resentment against Punjab in the drainage-induced water pollution disaster that has unfolded in the Manchar Lake, in Sindh.

Conflict over drainage and water quality in Sindh

Manchar was a source of rich fisheries and migrant bird hunting for the fisher communities, numbering about 50,000 until the 1990s. Most of these fishermen lived on boats in the middle of the lake and only came on shore for essential supplies. As one fisherman said, Manchar 'was their Dubai' that they never felt any need to leave. But as the MNV Drain started dumping agricultural and industrial effluents collected from a vast catchment including Balochistan, northern Sindh and southern edge of Punjab (Government of Sindh, 2015), the water became increasingly polluted leading to the collapse of fisheries, and aquatic flora, in addition to any other ecosystem services such as drinking and domestic water supply for which the lake had been a source for its communities (Ghaus et al., 2015). The estimated fisher population of the Manchar decreased from 50,000 living in 2000 boats to just 20,000 people. Migratory birds which were another source of protein for the lake residents collapsed from 25,000 birds counted in 1988 to 2,800 birds counted in 2002 (Zoological Survey Department, 2005). In addition, the lake was a source of irrigation through the Danister Canal and Aral wag canal to the lands surrounding the lake. Those canals too have turned poisonous and can no longer sustain agriculture, leading to the collapse of small-farmer livelihoods along those canals as well.

The local fisherfolk known as Mohanas as well as farmers were quick to follow plot line of the story often repeated for India in Punjab for the flood hazard. In the process, they enacted the national/inter-provincial scale in opposition to the local scale of the Manchar Lake, for example:

> Punjab imposed the MNV Drain on us. This place [Sindh] is run by Punjab. The military is Punjabi, even though my son is a captain in the army. But he has not turned into a Punjabi because I married him here in Sindh. (Farmer, Zamindar Bukhari, District Sehwan, 24/03/2015)

At the national scale of Pakistan, the military and Punjabi dominance is a familiar trope, and this trope was repeated to explain the local-level ecological collapse of the Manchar Lake and the accompanying misery that its fisher and small-farmer communities were suffering. But beyond the national versus local-scale conflict was also a profound understanding of the power asymmetries that were contributing to the state of the Manchar Lake, for example:

> There are a lot of factories and big landlords who dump their waste in the MNVD. We are poor (and) humble people. We cannot pick a fight with them and ask them to stop this to save ourselves. We just have to bear what they are dishing out to us. (Fisherman, Khan Mohammad Mallah village, 23/04/2015)

The ecological collapse has had multidimensional and catastrophic consequences, particularly for the fisher communities and spawned a new set of conflicts between the fisher and the farmer communities, as well as between the combined lake communities and outside middlemen, contractors and corporate interests. We shall outline some details of those conflicts later in this chapter.

Contrasting flood conflicts in Sindh and Punjab

As mentioned earlier, all irrigation diversion and storage infrastructure have a safe design capacity. Table 4.1 lists the safe design capacity of the barrages on the Indus. During flood season in the Punjab, the breaching sections of the levees are handed over to the Pakistan Army Engineers. The irrigation officers in the Punjab Irrigation Department have the authority to order the operation of the breaching section as soon as the safe design capacity of the infrastructure is exceeded. In the past, the officers were authorized to make that decision alone, but now there is a departmental panel that collectively decides the issue. However, there is some evidence to suggest that, in reality, senior officers can still take the decision on the spot (SE Punjab Irrigation Department, personal communication, 2015). The decision is subject to considerable outside pressure. The upstream farmers always want the breaching section to be operated to relieve pressure on their

Table 4.1 Control Station Design Discharge Capacities

River	Station	Design capacity (in ft^3/sec (cusecs))
Indus	Tarbela	1,500,000
	Kalabagh	590,000
	Chashma	950,000
	Taunsa	1,100,000
	Guddu	1,200,000
	Sukkur	900,000
	Kotri	875,000
Kabul	Nowshera	NA
Jhelum	Mangla	1,060,000
	Rasul	850,000
Chenab	Marala	1,100,000
	Khanki	800,000
	Qadirabad	807,000
	Trimmu	645,000
	Panjnad	700,000
Ravi	Balloki	225,000
	Sidhnai	150,000
Sutlej	Sulemanki	325,000
	Islam	300,000

Source: Sindh Irrigation Department.

lands, while the downstream farmers, particularly on the right bank, do not want it operated for the obvious reason of preventing flooding of their lands. The crux is that whichever local representative is more powerful or aligned with the government in power typically gets his or her way – within a certain envelop of the flood flows (see Mustafa, 2002). The irrigation department knows as well as the local residents that the safe design capacity number is a conservative one; where the advertised capacity is 175,000 cubic feet per second (cusec), as in the case of the Sidhnai Barrage on the Ravi, the structure can actually withstand up to 50,000–75,000 cusecs more than that. It is within that envelop, coupled with forecasts about additional rain or flood flows, that the presiding irrigation officer is subjected to pressures by different constituencies to (not) operate the breaching section. Historically, if the irrigation officer perceives any threat to the actual integrity of a structure, he will give the authority to the Pakistan Army Engineers to lower the explosive charges into their designated places along the levees and breach them, no matter what the pressure on them not to breach. It is no coincidence that no irrigation department in Pakistan has ever lost an infrastructure in floods, despite the invariably emotionally charged debates that take place over the operation of the breaching section. The debates take place as long as there is uncertainty about upstream conditions, weather forecasts and the structural safety envelop of the structure, but not after the safety envelop is exceeded.

Once the breaching section is operated, the general public, with some justification, invariably blames it upon the machinations of one powerful landowner or political leader or the other. It should, however, be clarified that powerful interests indeed influence the timing of the operation of the breach. Yet, there is no concrete evidence to suggest that any local landowner stores enough explosives to unilaterally blow up a levee on a mainstem river of the Indus Basin. Tools or bulldozers are patently insufficient or inapt for the task, given the size of the levees. The powerful interests also tend to heavily influence the operation of the breaching on canals that have to be breached to prevent wider destruction, as the following anecdote related by a senior officer of the Punjab Irrigation Department illustrates:

> My service was in DG Khan area. It is very hard to say no to politicians. In that district there are a bundle of politicians and they can straight away talk to the secretary or the Chief Engineers. In DG Khan the politician wanted to breach the canal at some point, to relieve the pressure of the hill torrent flood. The politician wanted me to breach at RD55, but I wanted to breach at RD66 [to prevent excessive damage to the canal]. I kept stalling him because I was convinced that

the canal will automatically breach at the spot that I was suggesting. Eventually the canal did breach at the spot that I had suggested and I just said that it is not my fault that God breached the canal at the spot that you did not want. (Senior Irrigation Officer, Punjab Irrigation Department, 27/04/2015)

One important event illustrative of the politics of the breaching section operation and the channelling of the flows took place during the 2014 floods. During these floods, a breaching section was operated upstream of Taunsa Barrage and the Muzzaffargarh Canal (Figure 4.3 illustrates the 2010 situation at the same location). The water, as it was flowing downstream, was headed towards the city of Muzaffargarh. In the words of a senior irrigation officer:

> In the 2014 floods, when the River Chenab had 600,000 cusecs, the breaching section at RD11/12 was operated. The upstream breaching was RD57 at Muzaffargarh branch. The water was likely to enter the city of Muzaffargarh city. There is a bypass in front of the city, and there were culverts in that bypass. We plugged the culverts [to save the city]. The upstream people wanted the culverts unplugged, while the downstream city people wanted them plugged. We closed the culverts to save the city. The same water started entering the Muzaffargarh branch and therefore relief was provided to the people who were sitting in the inundation zone. The whole area was ready for Rabi crop and we were able to allow the sowing. (Senior Irrigation Officer, Punjab Irrigation Department, 27/04/2015)

In this particular instance, a local member of the National Assembly (MNA) was arrested for manhandling local irrigation officers when they refused to unplug the culverts. The MNA was trying to protect the people in his constituency upstream of the culverts. These types of conflicts are invariably repeated all across the Indus River System, particularly in the Punjab.

To the discussion here, one must also add the conflict between irrigation and power generation imperatives of dam operations and flood management. At the moment, the standard operating procedures for dam operation in Pakistan require the chief engineers of all the water storage facilities in Pakistan to have their reservoirs filled on 20 August – the middle of the flood season. Inevitably, when floods exceeding the capacity of the dams reach the reservoirs, these have to be rapidly emptied to protect the dams, making bad floods worse – as happened in 1992 catastrophic floods when Mangla Dam was threatened in late September and it had to be emptied in a hurry to save it. There is an urgent need for flood management to be more fully integrated and prioritized in dam operations in Pakistan.

In Sindh, the situation is slightly different because the floods have been so infrequent that the irrigation department never specifically designated breaching sections upstream of the three barrages in the province. In the Sukkur Barrage area, for example, the Tori Bund upstream of the barrage was breached only once in 1976, and then too, at the behest of no less an authority than Zulfiqar Ali Bhutto, the then prime minister of Pakistan. The authority for breaching a bund in Sindh rests with the chief minister, none of whom, however, have ever exercised this power in the past. In 2010, the bund was breached naturally and vast swathes of territory on the right bank of the Indus, following the course of the MNVD, were flooded as a result. For the same reasons of the lack of drainage as in Punjab, the land remained inundated for a very long time and the drainage took several months.

There are, of course, unconfirmed reports of canal breaches by local bigwigs, but the mythology surrounding the breaching of levees invariably puts the onus on powerful politicians suspected of protecting their own. Whether that is true or not is beside the point. The point is that the opaque workings of the irrigation department, coupled with the deep public mistrust of the government and the powerful, lead the public to believe all sorts of theories about why breaches happen.

In the long run, breaches are ecologically and even economically a good thing for many of the reasons outlined earlier. All of Sindh had record harvests in the years following the 2010 floods because of the restoration of soil fertility and the recharge of groundwater. However, it is rather alarming that there are no designated breaching sections in Sindh as any infrastructural failure in the absence of clear operating procedures for the barrage operators would be catastrophic.

The flood hazard in Sindh and Punjab is further accentuated by arbitrarily built spurs and dams within the river bed that protects certain areas of land, often those belonging to influential people. All that spurs typically do is transfer the pressure of floods from one point to another through a perpetual cascade of worsening flood hazard along the banks of a river. This is another example of how conflict over flood plays out through competing influences on public institutions and the related expenditure of public money to protect one powerful interest at the expense of another.

The biggest issue in flood management in Pakistan is not whether breaching sections should be operated or not. It is more about transparency and early warning – and it is here that the government of Pakistan fails miserably at all levels. The state doesn't do too well with mediating conflicts arising out of its own engineered hazards like the ecological collapse of the Manchar either, as I shall outline later. In fact, if anything it becomes an enabling party in those conflicts against the weak and the poor.

Downstream conflicts from pollution hazard

The ecological collapse of the Manchar is structurally tied into the historical trajectory of irrigation development and subsequent Green Revolution in the agricultural sector. The Green Revolution high yielding varieties of crops are dependent upon enhanced inputs of chemical fertilizers, pesticides and water. The flat geography of the lower Indus meant that run off from agricultural fields will have to be drained through subsidiary and spinal drains to somewhere, with enhanced loads of toxic agricultural chemicals mixed in greater water flows. Beyond the collapse of fisheries and the floral life that the people depended on for nutrition is the pollution of drinking water sources – the aquifer underlying the Manchar Lake being the primary source. The government of Sindh has installed about twelve reverse osmosis (RO) plants in the Manchar area, which still leaves about 60 per cent of the populations around the lake dependent upon contaminated lake water for drinking and other domestic uses (interview with Mustafa Mirani 16/12/2015). In the village of Khan Muhammad Mallah (Figure 4.5), which is only accessible by boat from the shore of the lake, the water crises was particularly acute since the aquifer that the village depended upon had become contaminated. There were many diseases reported in the villages, as one of the women in a focus group discussion (Figure 4.6) stated:

> We get very sick because of bad quality water. My own child was for a few days and now my own grandson is very sick. I have had to pay PKR 6500 over the past week for his treatment, but the child is not getting well. (Woman, Khan Mohammad Mallah village, 23/04/2015)

In the village the RO plant had been stolen four days before our arrival. It is not an easy task to steal a RO plant and it requires considerable technical prowess. But even when the plant was operational, its location was highly problematic for the village residents because of caste and class conflict with the neighbouring farmer communities where the RO plant was located:

> The RO plant was installed far away on the mainland next to the road [they have to take a boat to the mainland and then walk about 200 metres to get to the plant on the side of the road side bund] because an influential person got

the government to install it on the mainland under the plea that others could benefit from it as well in addition to the fisherfolk. The reality is that this is a tribal area and if we had the plant here, we won't let outsiders to come here and take water either. So, we don't like going to the other tribe's area either for water. (Fisherman, Khan Mohammad Mallah village, 23/04/2015)

Solar RO plants have been provided to villages who have contacts with MPAs and waderas [landlords]. We would also like to have the solar plant, here in the village – if you could please help us get one. Other places have solar systems, which provide them electricity. Here at night I can't even see where I am going or if I am stepping on a snake. I have fallen in the water at times at night. Surely, we deserve at least that much. (Elderly woman, Khan Mohammad Mallah, 23/04/2015)

In other villages like Khair din Mallah the membrane for the RO plant has outlived its service life and hasn't been replaced by the government. It cost Rs.200,000 (approximately US$250 in 2015), something that the community could not afford. Under the circumstances fetching water had become particularly onerous for the fisher women who had to travel up to 13 kilometres into hostile tribal and farmer territory to fetch water and be harassed in the process of collecting water

Figure 4.5 A view of the Khan Mohammad Mallah Village, District Sewan.

Figure 4.6 Focus Group Discussion at Khan Mohammad Mallah village.

from contaminated sources (Figure 4.7). A bottle of mineral water cost about Rs.40 (equivalent to 50 cents at the time), which was beyond the reach of most households. Consequently, where women in the past were much more engaged in fishing and fish processing activities, now they are largely limited to managing water and other domestic duties.

Although the water conflict is at the local scale, its solutions are perceived to be at the provincial and regional scale by the local populace. While the drivers of the contamination hazard that they suffer are at the national/inter-provincial scale, the solutions are deemed to be in patronage from the provincial scale, for example:

> We have access to Syed Murad Ali [minister of finance at the time and now the chief minister of the Sindh province], who is a patron and like an elder brother to us. He is the local MPA. But the reality is that before we can get to him, we can be imprisoned by the police at the behest of a local big farmer, especially during election time. So, you see, it is not a surprise that humans have made it to the moon and Pakistan is still the same as before. (Fisherman, Khair Din Mallah village, 23/04/2015)

Comparable themes that feature in the domestic water supply conflict also feature in the land and water conflict at the local scale, borne of the same drivers as the conflict over domestic water, as I discuss in the following section.

Figure 4.7 Women scooping up brackish water from a groundwater source.

Big fish eats little fish: Farmers versus fishermen

The fisher communities had relative stability in their livelihoods before the decline of the Manchar Lake, though it was not a trouble free life. The fishermen were exploited by the contract system where contractors would hold a monopoly over the marketing of the fish thereby denying fair compensation to the fisherfolk. That problem was at least legally solved by the Sindh Fisheries (Amendment) Act 2011, whereby the contractor system was abolished in favour of a licensing system (licence fee being Rs.600/year or US$5). But even though the contractors for now have no legal right to obstruct fishing or marketing of fish in the lake, they continue to de facto interfere with the local fisher populations under patronage of the local large landowners. But beyond this ongoing conflict is the intensification of the conflict between fisher communities and farmers since the collapse of the lake ecosystem in the 1990s.

The fisherfolk–farmer conflict has its genesis in the administrative decision undertaken in 1976–7 on the behest of the then prime minister of Pakistan Zulfiqar Ali Bhutto. He supported the fisherfolks' demand to raise the water level of the lake so that the fisherfolk could have a buffer during the winter low flow years for fishing. The lake is supplied by various canals including the Aral

Manchar Canal, Aral Laki Canal and the Danister Canal in addition to the seasonal Gaj River. The raising of the lake level by the irrigation department meant that substantial portion of farming land was submerged. Although a compensation scheme was launched for the farmers, the process was marred by assorted claims and counterclaims and continues to be incomplete to this day (Syed Murad Ali Shah, finance minister of Sindh personal communication). The simmering resentment broke out into the open as the fisher communities were driven to the shores with the collapse of the lacustrine ecosystem, for example:

> Between farmers and fishermen we were all equal because we have our livelihoods and they have theirs. But ever since the fish catch has declined we have become like the servants of the *zamindar* [landholders]. (Fisherman, Khan Mohammad Mallah, 23/04/2015)
>
> We have a claim on the Manchar. [. . .] The big landlords got compensation, but no compensation to the ordinary people. Even our houses were where the Manchar is right now. In the past, with the sweet water, every time the water went down we would get a crop. We would do *rabi abadkari* [winter irrigation] when there was sweet water. (Farmer, Safi Thalo village, 24/04/2015)

The hunting and fishing rights in the lake were allocated by customary law in the past. People could hunt on their parcels of land and it was not material because fishers mostly stuck to the middle of the lake while the farmers hunted on the shore. Since the lake has been contaminated and the fisherfolk have been forced to migrate to the shore, the competition and resentment have been accentuated. But the conflict is also progressing towards the centre of the lake, with the entry of a Chinese firm doing oil exploration in the lake, for example:

> Since the MNVD started and the fish disappeared, the farmers and thekedars [contractors] have taken over the lake. Now, there is also this oil company that seems to own the lake, and not us. (Fisherman, Khair Din Mallah Village, 23/04/2015)

In late 2014 the Chinese geophysical services company BGP undertook a seismic survey, using explosives leading to considerable mortality among the low quality dhayya fish that is now populating the lake. The Clause 7 of the 1980 Sindh Fisheries Act specifically prohibits use of chemicals or explosives to catch fish or destroying aquatic life. But that prohibition evidently does not apply to oil exploration. To explain the disappearance of fish from the lake and the fish mortality, many interests including some newspapers started reporting on the fisherfolk using chemicals to catch fish (Baloch, 2015):

> When the water went bad for the first time, somebody said that because they throw in daal and mattar [lentils and peas], and sometimes DDT powder in the lake: that is why the water has gone bad. The farmers were blaming us for that. [. . .] We insisted that it was the RBOD. And about 300 people of ours were arrested for allegedly putting DDT powder in the water. Finally, the Sindh University proved that it was RBOD. (Mustafa Mallah village, 23/04/2015)

> Every time we catch fish by digging a hole we would not put chemical in there. In a 12 ft deep and 233sq. km lake, how much poison should one use to actually catch fish? Do you know how expensive chemical is? How much is fish worth? How could it economically make sense? Name me one example of anybody caught for putting chemicals in the water. They need to shut down the RBOD. (A fisherfolk, Sehwan 23/04/2015)

Even here the attempts to blame the fisher community are perceived by many of them in the context of inter-provincial-scale politics, for example,

> How can one do that [use poison to kill fish]? We can't find poison to kill ourselves, where will we find poison to kill the fish? [. . .] It is an accusation, and I think this is an accusation by the Punjab government. (Fisherman, Khair Din Mallah, 23/04/2015)

The farming communities, however, were quite positive about the oil exploration as they saw prospects for royalties from any oil that might be discovered and hence were more aggressively pursuing their claims on the lake bed, for example,

> If the BGP finds any oil in our land, because we have survey numbers for our underground property, we for sure will ask for royalties for the oil drilled from our land. (Farmer in Shafi Talo, 24/04/2015)

The fisher communities had a contradictory relationship to oil exploration. On the one had they want to supplement their meagre livelihoods by renting out their boats to the oil company. But even there they reported having to pay commissions to the large farmers and contractors to rent their boats. And on the other hand, they feared a permanent loss of the lake to oil exploration and the resultant contamination. The fisherfolk also resented the loss of their fish netting in the lake to speed boats used by the company. The fisherfolk tried to organize against the company, but were discouraged by the national-scale forces arrayed against them, particularly the paramilitary federally controlled Rangers:

> It was known that the federal government has sent them [the oil company], the Rangers were with them, and if anybody went against them, they would have

had an First Information Report (FIR – police report) against them and they will have gone to prison. . . . We are poor people we just get 1000 PKR for renting our boats. The farmers said to us that the waderas [big shots] would kill us [if we protested], and if not them then the Rangers will, so please don't spoil our livelihoods or lives. So, I thought if that's what the farmers want, I better step back. (Fisherman, Mustafa Mallah village, 23/04/2015)

The very local-scale fisherfolk–farmer conflict displays repetitive scalar politics at play. The national- and inter-provincial-scale analytical mind perceived drainage as a national-scale problem to be solved through mega-infrastructural projects like the RBOD. National priorities about oil exploration impinge upon local livelihoods and insert themselves into local-scale conflicts over domestic water, livelihoods and hunting rights. The local-scale subjectivities are about security of livelihoods and breaking out of the narrative of blaming the victims for their misery. The local fisherfolk and farmers are quite aware of the national- and provincial-scale forces at work and sometimes do engage in a conflictual discursive and material action against those forces – a stark manifestation of a hazardscape indeed.

Conclusion

The flood and pollution hazards are illustrative of multi-scalar forces that converge upon local life spaces and actors to define hazardscapes. The flood hazard in its present manifestation in the Indus is structurally linked to the design of the irrigation system. The priority for the system managers is to protect the irrigation infrastructure and that priority is met every time there is a flood event. The operation of breaching sections to protect infrastructure is a politico-technical process, where the populace gets to have an input into the process through its public representatives or patrons. But the loss of life and property inherent to the flood hazard can be mitigated through better warning systems, more transparent operations of the system and more radically, allowing the rivers room to flow, that is, by returning the areas where breaches are to be undertaken back to the river as wetlands, and providing flood-proof and flood-resistant infrastructure and livelihoods to the people. But such an undertaking will involve the type of democratic deliberative interaction between the system managers and the populace of the river, for which the present colonial bureaucracy is not very well suited (e.g. see Mustafa, 2002). The same could also be said of water pollution and drainage hazard. The farmer-oriented developmental vision of the system has a blind spot towards the fisherfolk livelihoods. The same vision is blind to how the so-called national-scale projects, like the RBOD, jeopardize local ecologies and livelihoods.

Is the conflict over the distribution of water-related hazards about or through water? The answer will have to be both. Materially the conflict is about protecting lives and properties from flood hazard or health and fisherfolk livelihoods from pollution. But the conflict is also through water between technocratic national-scale visions of development, technical progress and national integration, and local- and provincial-scale aspirations around identity, livelihoods and justice. The problem as illustrated by the case of floods and pollution hazard is about assuming a normal developmental trajectory where floods or water pollution are disruptions to be managed and isolated in the interest of maintaining course towards progress. For those living with those hazards, they are not just disruptions but the defining context of their existence. Re-centring the experiential hazardous view of water may yet be essential towards fairly negotiating the types of conflicts outlined in this chapter.

One of the first steps towards resolving rhetorical and actual conflicts over flood management is greater openness and transparency in the flood management institutions in Pakistan. These institutions, ranging from the Pakistan Meteorological Department (PMD) to the FFC, to WAPDA, to the provincial irrigation departments, are enthralled of the engineering/ technocratic paradigm of flood management. Their jargon-laden flood warnings and explanations are for the most part incomprehensible to the general public and even to educated members of society. For example, the assistant commissioner in Badin explained to me that in 2011 he received a fax warning from the PMD about a certain percentage probability of certain millimetres of rain falling in Badin over the next forty-eight hours. He had absolutely no idea what to make of that information, and hence did not do much beyond alerting his staff that something extreme was going to happen – without being sure what. The year 2011 saw one of the most devastating flood events in the district of Badin in lower Sindh. The flood managers have to be trained to translate their flood warnings and management plans so that they are understandable by the public and water managers as well. The media and the public also have to be educated in how the system works so that there is less opacity in the management and hence fewer unfounded conspiracy theories surrounding floods.

Second, people will always lobby the authorities for protection of their lands from inundation, as is their democratic right, even if they do it through the medium of patronage politics. All the same, forums need to be created for the public to articulate their concerns and for their concerns to be registered, while at the same time being educated about the compulsions of the flood managers.

Third, the designated inundation zones need to be prioritized for flood proofing, protection, rehabilitation, education- and warning-related initiatives. Particular focus should be on social vulnerability assessments to get some sense of who might be more vulnerable to flooding and how the government might offer them protection and rehabilitation in the aftermath of floods. The targeting of such inundation zones should be a special priority in terms of early warning and evacuation advice, which also need to be gender sensitive.

Fourth, post-flood drainage of inundation zones should be a priority in terms of helping people return to their homes and resuming their economic activity. The provision of pumps and undertaking of modifications in the infrastructure to facilitate drainage should be priority investments.

Fifth, flood protection spurs should be absolutely banned, unless there is a compelling case for protecting public infrastructure, such as hospitals, schools, grid stations, cultural monuments and so on. At the moment, they are simply a means to shift the hazard from one place to another, closely following the geography of power in society.

Sixth, the safe design capacities of the infrastructure should be reviewed specifically in light of the changes to be brought about by global climate change. Upgrading barrages and dams will be massive investments – but resources could be successively found for them if they are made a priority. In the same vein WAPDA and the PIDs could also update their outdated thresholds of low, medium and high floods given the longer time-series data available to them. The present thresholds are simply unrealistic and apparently haven't been updated in decades.

Seven, the public has to be educated that dams can have many benefits (and costs), but in the context of the Indus Basin, flood protection is not one of them. Few dams in the world, even if kept empty, could withstand the types of flood peaks that occur in the basin. Therefore, the polarized debates about dams, as per Chapter 3, need to be rationalized.

Lastly, conflict over floods is again deeply intertwined with issues of public trust and the relationship of the populace with the state. While the state seems to be enthralled of expensive infrastructure solutions, low-cost technical and financial assistance for flood proofing the properties of the poor could pay rich dividends in terms of preventing the loss of productivity and disruptions to people's livelihoods in the event of floods. Such steps will concomitantly help enhance trust between the state and the people.

Floods are a blessing and can be a curse. The fact that they are experienced more as a hazard is a social choice that society has made. It could also make a different choice, but for that to happen the voices of all, not just of the most powerful and vocal, will have to be heard. Rational interventions in the flood management field could lessen the prospects for conflict and ill will around the hazard, and possibly even foster a more cooperative ethos towards managing and living with floods.

5

Conflict over domestic water supply
The case of Karachi

With the 2017/18 'day zero' for water in Cape Town fresh in peoples' minds, in India in 2019, twenty-one cities are teetering on the edge of running dry. Chennai, a city built on marshlands and nestled in the subtropical monsoon region, is paradoxically the first one projected to run out of water. But few are posting such dramatic headlines for Karachi, a city in the desert. Therein is the hint that cities almost never absolutely go dry. The effects of water mismanagement borne of technocratic hubris and capitalist greed, like the drivers, tend to have differentiated effects. In Karachi as in other cities of the global South, water ran out for the urban poor a long time ago. The financial and physical effort of getting to water is a routine perversity for the urban poor. But this differentiated access to water and vulnerability to hazards from its non-availability are deeply intertwined, with geographies of class, ethnic and gendered conflict. Cities with their complexities of infrastructure, institutions and human diversity tend to be home to more concentrated examples of layered and cascading hazards – a defining characteristic of a hazardscape. With the fastest urbanizing population in Asia (Kugleman, 2013), the question of conflict over domestic water supply has immense implication for social peace and public policy in Pakistan. As in the preceding chapters I shall outline the axes along which water conflict actuates in Karachi and how social conflict is not just 'over' water but also 'through' water. The conflictual urban hazardscape of Karachi, I will argue, is also illustrative of the disconnect between the modernist technocratic gaze of power and how that gaze is contested and undermined to spawn the lived reality of the powerful and the weak in the city.

It is not a misstatement that Karachi is a mini Pakistan. Practically all of Pakistan's ethnic groups are represented in force in the commercial and industrial hub of the country that accounts for 35 per cent of the federal tax revenues of the country and 25 per cent of the GDP (Ahmed, 2012 and Zaidi, 2014). The city has also been suffering spates of ethnic and religious conflict, particularly over the last three decades. The water supply situation in the city is dismal to say the least. A very small proportion of water users – mainly those who live in affluent neighbourhoods – in the city have access to reliable and safe domestic water supply (Ahmed, 2014). The standard narrative of the water managers of the city is that there is an absolute scarcity of water in the city, where the demand is 1,000 million gallons per day (MGD) while the supply is less than 500 MGD (Kaleem & Ahmed, 2014). With such a framing of the problem, the solution is also quite clear: enhance the supply through large infrastructure projects. The story, as we have found out and as documented by many journalists, is much more complicated than that. This chapter is about illuminating the complexity of the geography of water and conflict over it in Karachi. The situation in Karachi as I outline in this chapter is not unique. To a lesser extent many of the themes I highlight in this chapter are hopefully relevant to other cities of the global South, such as the ones I mentioned earlier.

I will start with a physical and institutional outline of the water supply situation in Karachi in the following paragraphs. I will then proceed to highlight themes of de facto privatization of water in the city, and the role of social power in allowing differential access to water in the main body of this chapter. I will conclude with a discussion of the implications of water conflict in the city and some possible future pathways.

Physical and institutional context of water supply in Karachi

Karachi is a megalopolis of estimated twenty million people on the Arabian Sea at the eastern edge of the Indus delta. In the nineteenth century it was largely a fishing village on the left bank of the seasonal Lyari River. In 1883, the British colonial authorities dug shallow wells in the Dumlottee area along the banks of the Malir River. As the city expanded, post-independence additional water sources were tapped from Kinjhar Lake and Hub Dam on the Hub River in Balochistan. Because of substantial leakage and also non-revenue water, at the moment, about 20–25 per cent of the water physically leaks out of the system while 40 per cent of KWSB customers do not pay their bills (Kaleem & Ahmed, 2014 and Ahmed, 2014). This is not including water users who are simply not on the books of the KWSB. Historically different government departments have performed the function of water supply in Karachi. The KWSB was set up as a separate entity in 1996 under Karachi Water and Sewerage Board Act 1996.

The upscale housing authorities in the city have their own water supply arrangements: the military-sponsored Defence Housing Authority (DHA) is responsible for water supply to its residents, while Clifton Cantonment Board (CCB) supplies water to the residents of Clifton. All these elite entities get bulk water supply from KWSB, and unlike other low-income towns within KWSB's jurisdiction, which at times can only get maximum 30–40 per cent of their quotas, the military cantonment can get up to 133 per cent of its quota of water (Rahman, 2008). Lately, the posh housing societies have, however, also been hit by the water crisis, and their residents have had to rely on expensive supply through water tankers (Ousat, 2015).

The then Sindh CM Qaim Ali Shah initiated work on Greater Karachi Bulk Water Supply Scheme, also known as K-IV project, in 2015. To be completed over the next three years with a cost of over 25 billion rupees, the project was supposed to add 560 MGD to city's existing supply (Ghori, 2015); it was still incomplete in 2019. While this project is being promoted as a solution to the water problems of Karachi, as the following sections would illustrate, Karachi's water problems are much more complex than simply an issue of conflict over a scarce resource or of supply enhancement.

Privatization by other means: How it never did and can work for Karachi's poor

That the water access situation is dire in Karachi will be an understatement. There are neighbourhoods in Karachi, which have water supply infrastructures in place but haven't received running water for fourteen years. There are neighbourhoods where people drink brackish water as a matter of routine. Some of the more affluent neighbourhoods, for example the DHA, on the other hand end up getting more than twelve hours of water per day. Some of the key problems identified in the literature on Karachi water are as follows: insufficient supply, dilapidated infrastructure leading to 20–25 per cent line losses; insufficient revenue base for KWSB; corruption among the lower staff of KWSB who divert water to the highest bidder; illegal hydrants diverting water from residential customers towards water tankers; illegal tampering of water mains; and theft of water. Each one of the above-mentioned problem diagnoses is relevant, but even all together they don't explain the full water access picture in the city. But the most important paradigmatic issue with water supply in Karachi is state's abdication of its responsibility to provide domestic water to the residents of Karachi, and a faith that somehow private market will be able to make up the shortfall. We argue that the de facto privatization of water is at the heart of Karachi's hydro-hazardscape, with active illustrative contestations of that elite hazardscapes in the working class, Baldia Town, Orangi Town, Gadap Town and Liaqatabad Towns in the western part of the city.

There are officially 9 legal hydrants in Karachi, while unofficially Rahman (2008) estimated about 161 hydrants with overhead pipes for filling up tankers. There are more than 10,000 water tankers operating in the city. The tankers were introduced in early 1990s by KWSB as a stop-gap arrangement for plugging water supply gaps, until the network could be expanded, and water supply enhanced. Instead the tanker water supply has come to practically replace the piped water system. The cost of a 1,000 gallon tanker in Orangi town in 2015 was Rs.1,600 (1US$ = 100 PKR in 2015) closer to the pumping station and increased to Rs.3,000 and even Rs.4,000 further away from the water source. The tanker supply system is the de facto privatization of water in Karachi. In Orangi, for example, water supply to the area is pumped through two pumping stations: Germany pump and Disco pump. To these two pumps was added a third pump of Altaf Nagar, which was a real estate development north of Orangi established by the MQM, a mohajir[1] ethnic party. On the Germany and Altaf Nagar pumping stations illegal overhead pipes were installed to fill up water tankers, which became the main conduit for water supply in the neighbourhood (Figure 5.1). Since 2014 there has been no piped water supply to Orangi, and the water that does arrive at the Germany pump, closest to Orangi, is used entirely to fill up private water tankers. Until recently each tanker has to pay Rs. 400 to the reportedly MQM workers sitting there, to fill up, though since the MQM has gone underground after a military operation against it, reportedly many such overhead hydrants have been taken over the paramilitary Rangers who are responsible for maintaining law and order in the city.

The people of Orangi did not take the changes lying down, and in early 2015 they attacked and destroyed the pumping station at Altaf Nagar. It was somewhat surprising to hear that, since Orangi is a solid MQM stronghold where the party organization penetrates the lowest levels of the community. The following is the people's response to our query on how that happened:

> Water was here till about a year ago, but then it stopped since they brought the Altaf Nagar header on line. We took action ourselves and went about 3 times

---

[1] Mohajir is the local term used to refer to the descendants of the migrants from India at the time of the partition of the subcontinent in 1947.

Figure 5.1 Overhead pipe at the Altaf Nagar pumping station, Karachi.

to destroy it. Now, it does stand abandoned but they won't allocate the water from the mainline to the Germany pump that supplies the water to the Orangi neighborhood.

Q: How did you dare attack the Altaf Nagar Pump?

Ans: When there is a necessity people lose fear. [aside] Sometimes the lower level party people get together with the community to go against the higher ups. They know that the party can't go against the people, and they also know that upper leadership understands that they don't have a choice. (Community meeting, Orangi, 20/04/2015)

In this instance the people are producing the community scale through violent collective action. But it is not just about the additional pump but also about

Figure 5.2 Tanker trucks lined up to fill up at the Germany pump. This is the closest we could get to the pumping station safely.

the fact that their attempts at producing a community scale are undermined when the entire flow of water to the Germany pump is diverted to tankers when the water does arrive there. That same, stolen water is then sold through tankers to the households individually for whom that water was meant for in the first place – at times 10–100 times the price they would have had to pay otherwise.[2] This market mechanism then undermines the community scale by rescaling water supply to the household level. This rescaling is involuntary and violent. During the field work I visited many supposedly dangerous parts of the city, but never was more scared than when I visited the Germany pump. There were armed people keeping a watchful eye on the trucks as they were queuing to fill up. For health and safety reasons my research team chose to not get closer than 100 metres from the pump for the photograph in Figure 5.2.

In the absence of reliable water supply in Orangi, as in almost all the poorer neighbourhoods of Karachi, the quest for water is for all household members regardless of age (Figure 5.3). This de facto privatization of water, and its conveyance through private operators, contributes to the non-revenue water that KWSB staff blame for their inability to provide water, in addition to the lament about people not paying their water bills. But people do pay premium prices for water conveyed to them by tankers and water vendors, and they pay more than they would to KWSB. The issue here is not unwillingness or inability

---

[2] This is a rough estimate since KWSB's water charges are not volumetric but rather based upon the size of the property.

Figure 5.3 Little boys carrying water bottles from a water vendor in Orangi.

to pay, but rather state's privatization of water through its turning of blind eye to active water theft by the powerful.

Tankers are not the only way through which water has been privatized. There are a whole set of market relations around the manipulation of water valves to supply higher-paying customers in high-rise flats or more affluent neighbourhoods. In the neighbourhood of Gujjar Nala, for example, there hasn't been water for fourteen years, starting when high-rise flats were constructed down their water main, where the valves were manipulated to supply that new development on higher priority than the low-income neighbourhood.

> Up until 1990 there was enough water, but since the early 1990s there isn't any water here. There are flats down the water line from us, which do get water. I guess if they were to give us water, the flats won't get any water. In the past however, only the flats used to get water, but now there is an illegal area further down the line from the flats, which does get water as well now. See it is all about power. If there is an activist influential councillor he can get an extra line – truth be told: I got an extra line right here from the main line. (A focus group discussion member, Gujjar Nala, 24/04/2015)

Another way through which the ability to pay plays a role in access to water and conflict over water is through suction pumps. In Karachi, a plurality of households who can afford it maintains suction pumps on the premises. Therefore, the ability to access water crucially depends upon distance from the water main, the size of the pump and the ability to pay the electricity charges for operating the pump. Often, people with the higher ability to buy and operate more powerful pumps will divert water from the mains towards their households causing the people further down the system to get no water at all, as the following interviewee in the working-class neighbourhood of Gadap mentioned:

> We get 10 hrs of water every second day, but every time there is water, there is conflict. Everybody has a suction pump. So the people near the main pipe get water and the people further away do not get any water. . . . We, as the elders, ask people to shut off their pumps so that the people at the tail end can get water. But this is because we are here. In neighbourhoods where there is no arbitration or leadership, it inevitably comes to blows.
>
> [To his friend] Do you remember the day before yesterday, when there was loadshedding [scheduled outage of electricity to manage load for insufficient supply], everybody got water, little water with less pressure, yet everybody got it. (Resident Gadap Town 20/04/15)

It is this undermining of the community scale with an autonomous household scale through suction pumps that often drives community conflict over water, as the more powerful deprive their financially, or locationally, less well-off neighbours of even a small amount of water. In fact, every mode of water delivery, from tankers to piped water as manipulated by valves, differentially serves higher-paying customers better than those with lesser ability to pay. Precisely because market-based mechanisms are premised upon ability to pay, the poor are always going to be at a disadvantage, as it is, in fact, the case with the poor of Karachi. To the technocrats it's a supply issue, but this urban hydro-hazardscape follows the relief of power more than simple water supply. The result of this process is perpetuation of the household scale in conflict with each other and in negotiation with the technocratically preferred city scale. But the poor of Karachi don't quite take it passively; they do engage in collective action against powerful interests to reclaim their right to water at the community scale. It is to the power dimension of access to water that we turn in the following section.

Have power? Have water

The geography of access to water and conflict over water closely follows the fault lines of class, ethnicity and political affiliation in Karachi. The fieldwork for this report was conducted in the neighbourhoods of Bhagwandas, a Hindu and Sikh working-class neighbourhood in Baldia Town, Ghaziabad and Bilal Colony in Orangi Town, and Christian Colony in Gujjar Nala area in Liaqatabad Town. In each of these places the patterns of political patronage, antagonism, religious discrimination and gender politics intersect with class to spawn disturbing geographies of access and conflict over water. The case of Ghaziabad in Orangi town has been discussed with relation to the issues of power and exclusion centred around the Germany pump in the previous section. In the nearby Bilal Colony, comparable issues of occupation of water supply pumping stations by one political party or another come into play. This is the case, despite the fact that the neighbourhood is the heartland of MQM support in the city, and many young people of the area are activists who have been targeted by the recent anti-terrorism and crime operation of the Pakistan Rangers, initiated in March 2015. The operation has had visible effects on access to water as well:

> There is no water here, hasn't been for 6 months. There is a tap 200 m down from here, but we have to put in double motors of 1 hp each to get water for 2 hrs. In the past, water used to come but now all water is sold at the pump. There was a schedule before, it was politically decided even then, who will get water at what time. In lower Bilal, there was water but none in Upper [Bilal Colony]. My son would go to the donkey cart person and fill up water from him, who would fill it from the mosque in lower Bilal. We used to collect money to pay the valve man to open water for us. As long as there was Bhaia [the big shot of the neighbourhood] we could get water. Ever since he died there is no water. In the past there was patronage. Now, a lot of the people have died or gone underground, so we have no protection.
>
> We women sometime go the pump to ask for water, but they say where are your men? Send your men over and they can sit at the valve and make sure that you get your turn. But many of our men have been martyred [in the anti-MQM operation]. So where do we find men to send their way? Bhaia used to protect us, but he too is missing now. Now it is all the Rangers looking over the water situation. (Woman focus group respondents, Bilal Colony, 20/04/2015)

This description of the routinization of unspeakable violence is not separate from the state's national and international-scale politics about security, war on terror and Pakistan's perpetual geopolitical conflict with India. One of the key rationales given by the Pakistani state to undertake a violent operation against MQM (after having patronized it for most of the early 2000s) was that it was working on an Indian agenda to undermine Pakistan. Others like Anwar (2014), Anwar et al. (2019) and Mustafa et al. (2019) have touched upon this theme in considerable detail, and that argument doesn't merit detailed rehearsal here. The main point is that the state's production of global and international geopolitics ends up linking with the local-scale questions of water and crime in highly violent and gender differentiated ways.

In Bilal Colony, as in almost all of the other surveyed neighbourhoods, women are the main managers of domestic water, and hence the most concerned about its provision in their households. They often take a proactive role in trying to negotiate water delivery with either the donkey cart vendors, or even with the political party workers at the pump. Sometimes, which neighbourhood gets water at the pump can become a question of perceived allegiances of the residents of that neighbourhood to the party (MQM). Often, the politics of water come down to the level of discussing the ethnicity of the public official who laid down the infrastructure for water delivery, for example,

> When they set up the water pipes there was this boy Javed [X-en KWSB]. He is gone too. Actually, there was a fight once, and they stopped water then: I think it was because of that . . . [woman – who was quickly shut up by the MQM man].
>
> [Once the man leaves] The boy who set up the pipes was a Punjabi, perhaps that's what rankled the party. We went to the big man of the party and when he refused us water my daughter in law abused the party. At that the person got really angry and said: 'No water for you.' We apologized later and asked for forgiveness, but the man was adamant. So, we just threw stones, created a disturbance, and then came home [without water]. (Woman focus group discussion, Bilal Colony, 20/04/2015)

Whether women's reporting of MQM's (in this case) is the cause for them to be denied water is true or not is beside the point. In their perception, the party and the local toughs were the arbiters of their access to water, and it is to them that they had to go. Their conflict was not with the state, but with the local powerful interests for accessing water. This is also a case of not just conflict over water but ethnic and class conflict through water. This theme was to repeat itself in other neighbourhoods, for example, in Baldia Town where we had a street-side meeting with the local residents (Figure 5.4). Here, initially, the story was that every street and then houses have pumps to suck water from the mains towards the street water line. But upon further inquiry, it turned out that the neighbouring streets had highly organized residents from the Pashtun ethnic group who often saw to it that their valves were opened when there was water in the main and that their valve capacity was higher than others. Furthermore, they

Figure 5.4 Corner meeting with residents of Baldia Town.

also saw to it that other valves were forcibly closed to make sure that they get the water. As the resident ruefully declared,

> It all comes down to that we are weak and the people at other valves are strong. We simply cannot fight them. (Resident of Baldia Town, 21/04/2015)

Religious minorities in particular are vulnerable to being denied water, or for that matter even the right to live anywhere. In Bhagwandas Goth, for example, the largely Hindu and Sikh residents were even reluctant to refer to themselves by their religious name and instead simply used the neutral phrase: 'We, the minority people' (Figure 5.5). They had been moved multiple times since the 1980s when they had come into Karachi from Kandkot in interior Sindh. They have now been living in their present location, which was previously solid waste dump for the past twelve years. There have been attempts at moving them, but they have successfully resisted and have been able to enlist the patronage of the local PPP parliamentary representative to be left unmolested. One of the chief factors contributing to their vulnerability is their invisibility in the eyes of the state, because being illiterate and born in rural areas, many of them do not have birth certificates and are hence unable to gain a national identity card,

Figure 5.5 Community meeting in Bhagwandas Goth.

which is the proof of their citizenship. They therefore cannot access government services or gain any kind of employment in the formal sector, neither can their children get admission in schools and get an education. Consequently, they are exceptionally reliant on political leaders for patronage and protection against everyday harassments, which are common for the poor and the weak in Pakistan, especially if one is from a minority community, for example,

> I was pulled over by policemen at night in Karachi. They took away my wallet and my mobile and told me to get lost. I said you guys are being unfair and I have contacts with the MPA [specifically MPA Lalu Khet] who you will have to deal with. They rebuked me by saying: 'you pimp of a Hindu you talk back to us?' I said: 'look in my mobile phone. If you find the MPA's number in there give me back my stuff, if you don't then I will walk away or you can arrest me.' They did look up the number and saw the MPA's number. So, they quietly handed me back my mobile and the wallet and let me go. (A Bhagwandas resident, 21/04/2015)

The extreme dependence upon patronage is also apparent in the water sector. The local MPA sanctioned a water supply scheme for the residents. The infrastructure for the water supply was laid out as was the sewerage system, but there has never been any water in any of the pipes (Figure 5.6). The dependence of the residents on water tankers is complete, and a major financial burden. There was a water tanker delivering water at the time of my visit. The water that the tanker delivered is shown in Figure 5.7. The water was clearly unfit for human consumption and was, in fact, brackish. The hazard of sociopolitical marginalization cascades into state violence, which cascades into access to expensive brackish water and the accompanying health hazards. To the residents, their 'minority' status and poverty is intimately linked in this hazardscape to their exposure to water scarcity. The quality of water being consumed by the residents of Bhagwandas is no different from that being consumed by the residents of Christian Colony in Gujjar Nala. There one resident quite bluntly said:

> Simple thing is, if you have himat [strength] there is water. No himat, no water: keep buying tankers of saline water and drink away that swill. (Resident Gujjar Nala 23/04/15)

In Gujjar Nala, there are serious intra- and inter-community conflicts bearing upon access to water in the community. At the intra-community level, while we were sitting on the road side having a conversation with the few community

Figure 5.6 Sewerage line manholes in Bhagwandas.

Figure 5.7 Water delivered by a tanker in Bhagwandas. This water is used for drinking as well.

representatives, about five to six children and adults knocked on the door of one of the persons sitting with us, across the street from where we were sitting (Figure 5.8). As one of the participants in the discussion said:

> You see that little girl there [she was the 3rd one to knock while we were sitting there] she is not going to get any water. Giving away water is a problem [because one doesn't have enough]. Not giving water is also a headache and cause for conflict and fights [because people say that you are being stingy with water and not being a good neighbour]. (Focus group discussion member, Gujjar Nala, 24/04/2015)

The lanes of the neighbourhood are too narrow for tanker trucks to get in. The influential community member on the main road has had an illegal connection from the water main connected to his house. The rest of the people have to buy water from water vendors, or tankers on the main road, or from people who have borewells of saline water on their property, and then carry it in carts into the neighbourhood. The quality of drinking water here is no better than in Bhagwandas. The reason they have not had water for a long time is not only because of the apartment complex that is down the pipe from where they are but

Figure 5.8 Girls knocking on the door to request water.

also because of their conflict with the neighbouring Pashtun and other Muslim communities, as narrated in their own words below:

> In the 1990s the Pathans [Pashtuns] up the 36 [24 inch according to KWSB] inch main water line [passing by the Christian Colony of Gujjar Nala] started closing the valve to direct more water to their neighbourhood. We went out there and fought them, but they had guns and TTs and we can't put our children's life at risk. In any event, Master Niaz finally got the valve opened, but once master Niaz (a local tough) was gone, they closed it again. Finally, the Pathans just opened the valve a few rings up and poured concrete in it, so that it can't be tampered with, and limited flow does come down the line towards us. But our 4 inch pipe for the neighbourhood is connected to the top of the pipe, so the water level in the main 24 inch pipe never really reaches high enough to get into the authorized 4 inch pipe. My other 4 inch pipe that I got laid, however, is connected to the bottom, so it does get some water every ten days or so.
>
> But beyond the valve there was this person Lala Yaqoob, who won't let us open our line to the community. There are all these Muslim neighbourhoods on the other side of the main line, which have also hooked in connections to the main. Lala Yaqub was the gangster here. We complained to the MQM unit in charge. He said, come with me. When we got to Lala Yaqoob, Lala tried to

talk tough to the MQM guy. He [MQM man] said to him, speak very softly to me, you have no idea what I can and will do to you. He scared Lala Yaqoob into laying off our line and letting us open the valve to our line. (Focus Group Discussant, Gujjar Nala, 24/04/2015)

The largely Christian community in this Gujjar Nala neighbourhood is staunchly pro-MQM, because it is through MQM's largely secular politics that they get the necessary patronage to fight off competing claims on resources, including water by other communities. The other communities are often not averse to playing the religious card to intimidate minority poor communities, such as the Christians of Gujjar Nala. For this community, as for many others in Karachi, there is perpetual competition to influence the KWSB employees to preferentially open their water supply valves, and negotiation of ethnic or religious vendettas through water. The highly violent politics of Karachi is also experienced through water for these residents, for example, as one of the community leaders said:

We have not raised our children to get into fights. They killed my young son: why? Because I raised a voice for my community and they killed him to put me down. There is water all around us, but no water for us. . . . Reality is that if Bhaia [the local MQM tough] helps us we get our way, but we need his protection. The network was for us, but now the Muslims also take water from our 4 inch line and their connections are completely illegal. With those Muslims there are always fights and there is some sector in charge or unit in charge extorting money for mediation or backing one side or another. If we want to take our rights, we have to sacrifice. (Christian community leader, Gujjar Nala, 24/04/2015)

The dominant narrative on the water situation in Karachi emphasizes the tanker, land and water mafias, and with some truth. The official version is about paucity of water where the K-IV water infrastructure project will provide all the relief. But beyond these meta-stories are the nitty-gritty details of how the poor and the vulnerable make alliances and fight every day, just to get access to water in a highly politically, ethnically and socially polarized city. The solutions to Karachi's water problems and conflicts are going to have to move beyond the standard narratives and basket of solutions. We turn to that issue in the concluding section of this chapter.

Conclusion: Possible ways forward

As the foregoing discussion illustrates, Karachi's water conflicts map on to the ethnic, religious, class and political conflicts in the city. Concerns with supply enhancement, system upgrade and revenue generation are fair enough and well documented and are the building blocks of the technocratic production of Karachi's hydro-hazardscape. But beyond those, there is a need for a paradigm shift from the technocratic visions of city and regional scale vast, networked piped system serving household scale water users towards more local and community scale modular, possibly stand pipe-based systems for the poorest. Pouring money into sewerage infrastructure, for which there is no water to operate, and providing water connections to individual housing units, which never get wet, as a mode of patronage is dysfunctional. The same money could get used to dispense water from central storage tanks and stand pipes so that people can at least have some access to quality water, instead of paying exorbitant prices for very low quality water. The production of the city scale through networked water and megaprojects in the technocratic sense papers over the local-scale politics of class, ethnicity and gender. The local-scale patronage politics keeps the subaltern on the defensive, while the oppositional politics of the subaltern link questions of water access to the city- and national-scale themes of identity, citizenship and rights.

Second, the governance of water also has to be devolved to the local scale with a rights-based approach to be enforced by the provincial-scale authorities. While devolution might be effective in terms of facilitating public demands and effective administration of the system, it will still have to have oversight for quality control, and ensuring that right to water is respected. In this vein, closer links between a representative city government and water administration will be a step in the right direction.

Third, the water conflict in Karachi as in other urban areas of Pakistan has to be viewed in its proper political context. City- and local-scale politics over and through water cannot be wished away; they can only be managed. The communities of Karachi are already well organized, but towards negotiating ethnic and class politics, mostly originating at the national scale. The national-scale ethnic labels tump local-scale potentialities for alliance building and class solidarity. The question that arises then is whether the local-scale organization's strengths can be mobilized to negotiate a rights-based approach, instead of the ability-to-pay-based approach, to water access with the city authorities. That will involve re-centring the subaltern view of the interconnectedness of the issues of political patronage, violence, ethnicity, gender, social power and water access in the city's hazardscape.

Lastly, there is no alternative to the state at the provincial and city scale playing its role as a provider of basic services like water in urban areas. The rich may be able to insulate themselves from the day to day privations suffered by the vast majority of the residents of cities like Karachi. But they cannot always remain insulated from what is happening to their fellow citizens. Politics recast as something not just about electricity, roads and lack of corruption but rather about redistributive justice, basic rights, identity and citizenship will create the necessary bridges for citizens of urban areas to empathize, communicate and hopefully make common cause with others with very different backgrounds and social status than their own. The hydro-hazardscape of Karachi is highly politicized and it should be. The question is, what is the type of politics that animates it? If it is classist politics intersecting with ethnic and gender divisions, then the consequence is what is discussed in this chapter. But if those scalar politics can be redirected towards linking local with the city scale for place-based water rights, then the outcomes could be diametrically different to the realities of urban water conflict at present.

6

Conclusion

Unlike the overwhelmingly international transboundary focus of the scholarly literature on water and conflict, I have added my voice to the nascent literature on sub-national hydropolitics through an illustrative study of water conflict in Pakistan. I have reviewed case study of meso-scale inter-provincial water conflict, conflict around flood and pollution management at the inter-provincial and local scale, conflict between fisher and agricultural interests at the local scale, local-scale conflict at the canal command and village water course level and finally again local-scale water and conflict in Karachi as illustrative of urban water and conflict. In this concluding chapter I will review the key insights of each of the chapters and then conclude with reflections on the conceptual insights that can be drawn from the cross-scalar approach, and the empirical evidence presented for each of the case studies.

The inter-provincial water conflict, on the one hand, is certainly about the material supplies and distribution of water in the Indus Basin between the upper and lower riparian provinces. On the other hand, water also becomes a medium through which discursive contestations about identity, ethnicity, development and nature of the Pakistani polity are negotiated. Is the lack of trust between Sindh and Punjab driving the water conflict or is the water conflict causing the lack of trust? There is no way to definitively answer that question one way or the other. The chapter on inter-provincial water conflict, however, argued that there are structural compulsions endemic to the upstream/downstream riparian dynamic, and the highly regulated nature of the Indus Basin that drive the conflict over water distribution between provinces. The controversy over KBD, however, is more about competing visions of development, and nation-scale versus local- and regional-scale politics, and problem definitions.

The national-scale expert view perceives an existential hazard of water stoppage by India and the doomsday scenarios of glacial melt and water scarcity to argue for mega surface storage projects like the KBD. The national-scale production of hazards is contested by regional/provincial-scale pleas for ecology of the Indus delta, and livelihoods of the Sindhi farmers and fisherfolk. In a self-similar pattern across the international and sub-national scale, Sindh too fears stoppage or curtailment of water to it by Punjab if it were provided the infrastructural means to do so, just as Punjab fears the water stoppage by India. To the technocrats at the national and the provincial scale, the inter-provincial hydro-hazardscape is informed by disruptions to the dominant average-based model of water availability. The oppositional interventions of the activists and politicians in the hazardscape are not about destabilizing the hegemony of the average-based models of water flow but instead about place making at the provincial and local scale. The inter-provincial water conflict does echo the theme of sectional identities as informing water conflict, developed by Moore (2018). But unlike Moore (2018) I maintain that such mobilization is not an aberration, but rather inevitable, and even desirable, to contest the power relations productive of the national scale in hydropolitics. The chapter on inter-provincial water conflict illustrates that it is not decentralization that drives water conflict as per Moore (2018), but rather the imposition of the national scale over meso- and local-scale subjectivities. Decentralization may provide outlets for articulating and negotiating conflict, but centralization doesn't make the conflict disappear; it simply makes it invisible, for example, the case of One Unit in Pakistan and the signing of the IWT as a result.

In the same vein as inter-provincial hydropolitics, conflict over flood management is also framed as an issue of technical management of flood peaks in the Indus system, on the one hand. On the other hand, floods too are imbricated with meaning about provincial rights, ill intent of the upper riparians at the national and international scales and the power politics at the local scale. Given the nature of the humanly transformed hydrology of the Indus, there is nothing natural about flood frequency and intensity in the basin. Which places are flooded and when are contingent human decisions pregnant with power relations, not just contemporaneously but since the inception of the system. Similarly, drainage and pollution issues are analytically framed as infrastructural problems to be addressed through mega-infrastructural projects like the MNV Drain. The waterlogging and salinity, as well as industrial pollution, are hazards that have resonance at the national level, and they attract national and international-scale investment. The local-scale subjectivities, health and livelihoods intertwined with ecology are occluded in that framing. The fractal nature of scalar conflict and the contestations within the hazardscapes characterized by multiple cascading hazards and power relations are put starkly in relief through the conflict over floods and water quality.

At the local scale of conflict over surface and groundwater in watercourses and Karezes respectively, I highlighted how the national and international scales act upon the local scale to precipitate conflicts over and through water. For Karez irrigation, the international-scale donors, ignorant of the local-scale multiple values of the Karez system, pushed for replacement of an ecologically and socially sustainable system with the 'demand-based' tubewells. The result was the virtual demise of the Karez system and the rise of water conflict where none was known before. The imposition of national-scale agricultural production imperatives upon the local-scale water management system, which was predicated upon cooperative management to negotiate seasonal scarcity, is one of the drivers of water conflict in Balochistan. The other driver is the techno-centric recipe for drought resilience through tubewells. The Baloch hydro-hazardscape features new water conflicts driven by scalar power/knowledge and political economic relations as well as analytical techno-fetishes.

In case of local-scale surface water, the evidence seem to point to the fact that water generally becomes an effective weapon for individuals and communities to play out their disagreements and conflicts about other issues, for example, political allegiances, family and caste conflicts and so on. Many such conflicts are embedded in national-scale partisan politics. State in this context often plays the role of an enabler for more powerful actors through its acts of omission, that is, by ignoring illegal deprivation of water to tail-end water users; or even commission, by simply handing over formal control to large farmers, as in the WUAs. Here the role of post-colonial Pakistani state has to be reoriented towards equitable water delivery instead of the practice of the 'science of the empire' that is its inheritance (Gilmartin, 1994). But while the recasting of the role of the state remains a larger strategic objective, it may be noted that the role of the irrigation infrastructure as an anti-drought resilience measure remains at the heart of the expert view of the surface irrigation system. The reality of the system, however, is more about commercial cash crop production and differential exposure of smaller and tail-end farmers to the twin hazards of water scarcity and floods (Mustafa, 2002). How the local-scale conflict unfolds – typically in favour of the larger farmers – is largely a function of the disconnect between the expert and experiential realities of the hazardscapes, in addition to the scalar power politics linking the state and the local water users.

The chapter on Karachi documents how conflict over water is along the ethnic, class and religious fault lines in urban Pakistan, especially Karachi.

On the one hand, class-based conflict is because of the de facto privatization of water in urban Pakistan. The commodification of water precipitates conflict along class lines, where people unable to afford higher prices for water through tankers or through suction pumps have to organize and sometimes violently take matters into their own hands. On the other hand, the powerful steal water from, or ration water to, the weak. The weak have to gain patronage of political parties and at times gangsters to assert their claims on water or to appropriate it from others. Here again the self-similar pattern of the distribution of water scarcity at the household and neighbourhood level is reminiscent of the regional and inter-provincial scale hydropolitics.

At the heart of water conflict in Karachi is the basic infrastructural artefact of networked, piped water system that purports to produce the scale of the city. It is at that scale of the city that KWSB has jurisdictional authority. But that authority is fragmented by the existence of more powerful jurisdictions, for example, the military Cantonment Boards, Defence Housing Authorities or more elite upscale sub-divisions, to which KWSB becomes a bulk supplier. Even within the poorer Karachi Municipal Corporation jurisdiction, there are de facto inequities based upon class and ethnicity. The class-, ethnicity- and religion-based water conflict is enacted through suction pumps at the street level, and through bigger pumps on water mains and water tankers at the neighbourhood scale. But equally, the conflict over water reproduces class, ethnicity and religious fault lines, thereby undermining any prospects for progressive alliance building for claim making upon the state and the civil society. The approach towards hydropolitics in cities like Karachi has to be both local-infrastructural and overtly political. The regional-scale mega projects, although relevant, are likely to be of limited efficacy in addressing water scarcity, quality and conflict in megacities like Karachi.

To sum up, water and conflict are related along two registers – conflict over water and conflict through water. The evidence has been mixed in terms of defining the causation of water and conflict. At the inter-provincial level of water and flood management, the conflict is indeed about water. The water infrastructure debates also feature disagreements about national versus local-scale development and conceptions of polity. At the local scale the water conflict over irrigation water or between fisher and agricultural interests is mostly conflict through water about development, land, power and resources. In urban Pakistan the water conflict is certainly about water, but then that conflict closely follows the contours of class and power relations.

The object of all accounts of hydropolitics at international and the nascent sub-national scale has been to understand the contestations over 'normal' water distribution. A key argument of this book has been that normality is a cultural artefact of modernity, which papers over the uncertainties and hazards that have characterized human–environment relations across time and space. The normality-based predictive models may have legitimized and perhaps even enabled the accumulative social systems that came to be under modernity, but those systems in turn have generated human-based instabilities and mega-uncertainties like the climate change. The disconnects between the hazardous and normal/developmental views of the world are the key drivers of water conflict across scales, as the cross-scalar accounts of water conflict in this book illustrate. Conflict in itself is not bad – unjust or coercive cooperation is worse. I have argued that re-centring the hazardous view of human–environment relations in hydropolitics provides the analytical lens that is closer to the experiential reality of a hazardous world for the subaltern and the weak, and hence more conducive towards representing their subjectivities and interests in cross-scalar hydropolitics. Water and conflict will exist and they should. The question is, how can more emancipatory and just outcomes be ensured in any water conflict? Hazardscapes perspective, I argue, is one way of working towards that.

Water is life, and life is embedded in politics and culture. It is essential that politically and culturally mediated values of water beyond its simple material value are acknowledged in water management. Water is unique because there is no substitute for it, and it is saturated with all the aesthetic, symbolic, material, cultural and political impulses that humans bring to life. Like life water management must also take into account all of these impulses. These impulses are concretized in the human life spaces and geographies at the local scale. Therefore, I endorse local- and regional-scale hydropolitics, which are relevant to lived life spaces of water users. The nation state scale is a historically contingent abstraction that cannot trump the right to water, livelihoods and ecology at the local scale. Much of the water conflict is a manifestation of people's attempt at claiming legitimacy for their local-scale life spaces, against the encroachment of national-scale imperatives. Democratic co-production of geographical scales, from local to national to global, may yet be the biggest moderating influence on water conflicts in Pakistan or anywhere.

References

Afzal, H. L., 1995. Settling Disputes between Ethno-Regional Groups in Young Democracies: Distributing the Indus Waters of Pakistan. PhD diss., University of Michigan.

Ahmed, A., 2012. How important is Karachi to Pakistan?. *The Business Recorder*, Retrieved on 28 December 2015 from URL: http://www.brecorder.com/weekend-magazine/0:/1186182:how-important-is-karachi-to-pakistan/?date=2012-05-05.

Ahmed, A., 2014. Troubled Waters: What Makes Water a Mismanaged Commodity in Karachi. *Herald*, 54–70.

Aijaz, A. & Akhter, M., 2020. From building dams to fetching water: scales of politicization in the Indus Basin. *Water*, 12, 1351. https://doi.org/10.3390/w12051351.

Akhter, M., 2015. The Hydropolitical Cold War: The Indus Waters Treaty and State Formation in Pakistan. *Political Geography*, no. 3: 65–75.

Akhter, M., 2015. Infrastructure Nation: State Space, Hegemony, and Hydraulic Regionalism in Pakistan. *Antipode: A Radical Journal of Geography*, 47, no. 4: 849–70.

Akhter, M., 2016. Desiring the Data State: Hydrocracy and Depoliticization in the Capitalist Periphery. *Transactions of the Institute of British Geographers*. doi: 10.1111/tran.12169.

Akhtar, M., Ahmad, N. & Booji, M., 2008. The Impact of Climate Change on the Water Resources of Hindukush–Karakorum–Himalaya Region under Different Glacier Coverage Scenarios. *Journal of Hydrology*, 355, no. 1–4: 148–63.

Alam, R., 2019. *A constitutional history of water in Pakistan*, Islamabad: Jinnah Institute.

Alam, U. Z., 2002. Questioning the Water Wars Rationale: A Case Study of the Indus Waters Treaty. *Geographical Journal*, 168, no. 4: 341–53.

Ali, I., 1988. *The Punjab under Imperialism*. Princeton, NJ: Princeton University Press, 885–1947.

Allan, S., 2011. Introduction: Science journalism in a digital age. *Journalism*, 12, no. 7: 771–77.

Anwar, N. H., 2014. The Bengali can Return to his Desh but the Burmi can't Because he has no Desh. In: M. Baas, *Transnational migration and Asia*, pp. 157–178.

Anwar, N., Sawas, A. & Mustafa, D., 2019. 'Without Water there is no Life': Negotiating Everyday Risks and Gendered Insecurities in Karachi's Informal Settlements. *Urban Studies*, 1–18. doi: 10.1177/0042098019834160.

Archer, D. R., Forsythe, N., Fowler, H. J. & Shah, S. M., 2010. Sustainability of Water Resources Management in the Indus Basin under Changing Climatic and Socio-economic Conditions. *Hydrology and Earth Systems Science*, 14, no. 8: 1669–90.

Baloch, S., 2015. The Murky Waters of Manchhar Lake. *Dawn*. Retrieved on 13 December 2015 from http://www.dawn.com/news/1159661/murky-waters-of-the-manchhar-lake.

Bandaragoda, D. J., 1999. *Institutional change and shared management of water resources in large canal systems: results of an action research program in Pakistan*. Colombo: International Water Management Institute: RR36.

Bates, S. F., Getches, D. H., MacDonnell, L. J. & Wilkinson, C. F., 1993. *Searching out the Headwaters: Change and Rediscovery in Western Water Policy*. Washington DC: Island Press.

Blomley, N., 2006. Uncritical Critical Geography? *Progress in Human Geography*, 30, no. 1: 87–94.

Brochmann, M. & Gleditsch, N. P., 2012. Shared Rivers and Conflict—A Reconsideration. *Political Geography*, 31, no. 8: 519–27.

Chatterjee, P., 1983. More on the Modes of Power and the Peasantry. In: R. Guha, ed. *Subaltern Studies*. New Delhi: Oxford University Press, pp. 311–49.

Dalin, C., Yoshihide, W., Kastner, T. & Puma, M. J., 2017. Groundwater Depletion Embedded in International Food Trade. *Nature*, no. 7647: 700–04. doi: 10.1038/nature21403.

Ghaus, K. M, M. A I, M., A, N., A, N. & A, T., 2015. *Gender and Social Vulnerability to Climate Change*. Karachi: Social Policy and Development Centre.

Ghori, H. K., 2015. Work Begins on Rs. 25.5 bn K-IV Water Project. *Dawn*. www.dawn.com/news/1187431.

Gilmartin, D., 1994. Scientific Empire and Imperial Science: Colonialism and Irrigation Technology in the Indus Basin. *Journal of Asian Studies*, 53, no. 4: 1127–49.

Gleditsch, N. B. M., 2012. Shared Rivers and Conflict – A Reconsideration. *Political Geography*, 31, no. 8: 519–27.

G.P., 2005. Government of Pakistan. Zoological Survey of Pakistan 2005. Zoological Survey Department.

Gulzar, F., 2018. Water Crisis: Pakistan Running dry by 2025, Says Study. *Gulf News*, 1. https://gulfnews.com/world/asia/pakistan/water-crisis-pakistan-running-dry-by-2025-says-study-1.2230115.

Henderson, G., 2003. What (else) we Talk about When we talk about Landscape: for a Return to the Social Imagination. In: C. Wilson & P. Groth eds. *Everyday America: Cultural Landscape Studies after J. B. Jackson*. Berkeley: University of California Press, pp. 178–98.

Hewitt, K., 1983. *The Idea of Calamity in a Technocratic Age*. Boston, Mass: Allen and Unwin.

Hoefle, S. W., 2006. Eliminating Scale and Killing the Goose that Laid the Golden Egg?. *Transactions of the Institute of British Geographers*, 31, no. 2: 238–43.

Homer-Dixon, T., 1999. *Environment, Scarcity and Violence*. Princeton, NJ: Princeton University Press.

Hulme, M., 2009. *Why we Disagree about Climate Change: Understanding Controversy, Inaction and Opportunity*. Cambridge: Cambridge University Press.

Jacobs, J. & W, J., 1994. Flood-hazard Problems and Programmes in Asia's Large River Basins. *Asian Journal of Environmental Management*, 9, no. 2: 96–108.

Jonas, A. E. G., 2006. Pro Scale: Further Reflections on the 'Scale Debate' in Human Geography. *Transactions of the Institute of British Geographers*, 31, no. 3: 399–406.

Kaleem, M. & Ahmed, M., 2014. Tapping into Trouble. *Herald:* 44–53.

Karpouzoglou, T. & Vij, S., 2017. Waterscape: A Perspective for Understanding the Contested Geography of Water. *WIREs Water*, 4, no. 3: e1210.

Kates, R. W. & Burton, I., 1986. *Geography, Resources and Environment*, Vol. 2. Chicago: University of Chicago Press.

Khan, H., 2009. *Constitutional and Political History of Pakistan*. Karachi: Oxford University Press.

Khan, I., 2014. Pakistan Will Ask India to Inform before Releasing Dam Water. *Dawn*. Retrieved on 6 July 2020 from URL: https://www.dawn.com/news/1116447.

Kiani, K., 2015. IRSA questions 'closed basin' description of Indus by lenders, *Dawn*. Retrieved on 6 July 2020 from URL: https://www.dawn.com/news/1227820.

Kugleman, M., 2013. *Four Key Challenges for Pakistan's New Government*. IPI Global Observatory.

Laghari, A. N., V, D. & R, W., 2012. The Indus Basin in the Framework of Current and Future Water Resources Management. *Hydrology and Earth System Sciences*, 16, no. 4: 1063–83.

Latour, B., 2005. *Reassembling the Social: An Introduction to Actor-Network Theory*. Oxford: Oxford University Press.

Leitner, H. & Miller, B., 2007. Scale and the Limitations of Ontological Debate: A Commentary on Marston, Jones and Woodward. *Transactions of the Institute of British Geographers*, 32, no. 1: 116–25.

Linton, J., 2010. *What is Water? The History of a Modern Abstraction*. Vancouver and Toronto: UBC Press.

Marston, S., Jones, J. P. & Woodward, K., 2005. Human Geography Without Scale. *Transactions of The Institute of British Geographers*, 30, no. 4: 416–32.

Menga, F. & Swyngedouw, E., 2018. *Water, Technology and the Nation State*. Routledge, p. 226.

Messerschmid, C. & Selby, J. 2015. Misrepresenting the Jordan River Basin. *Water Alternatives*, 8: 258–279.

Michel, A. A., 1967. *The Indus Rivers: A Study of the Effects of Partition*. New Haven, CT: Yale University Press.

Moore, S. M., 2018. *Subnational Hydropolitics: Conflict, Cooperation, and Institution-Building in Shared River Basins*. New York, NY: Oxford University Press.

Mustafa, D., 2001. Colonial Law, Contemporary Water Issues in Pakistan. *Political Geography*, 20, no. 7: 817–37.

Mustafa, D., 2002. Theory versus Practice: The Bureaucratic Ethos of Water Resource Management and Administration in Pakistan. *Contemporary South Asia*, 11: 39–56.

Mustafa, D., 2005. The Terrible Geographicalness of Terrorism: Reflections of a Hazards Geographer. *A Radical Journal of Geography*, 37, no. 1: 72–92. https://doi.org/10.1111/j.0066-4812.2005.00474.x.

Mustafa, D., 2007. Social Construction of Hydropolitics: The Geographical Scales of Water and Security in the Indus Basin. *Geographical Review*, 97, no. 4: 484–501.

Mustafa, D., 2013. *Water Resource Management in a Vulnerable World: The Hydro-Hazardscapes of Climate Change*. New York, NY: I.B. Tauris & Co. Ltd.

Mustafa, D. & Wescoat, J., 1997. Development of Flood Hazards Policy in the Indus River Basin of Pakistan, 1947–1996. *Water International - WATER INT*, 22: 238–44. doi: 10.1080/02508069708686712.

Mustafa, D. & Qazi, M., 2007. Transition from Karez to Tubewell Irrigation: Development, Modernization, and Social Capital in Balochistan, Pakistan. *World Development*: 35, no. 10 1796–1813.

Mustafa, D. & Sawas, A., 2013. Urbanisation and Political Change in Pakistan: Exploring the Known Unknowns. *Third World Quarterly*, 34, no. 7: 1293–1304. doi: 10.1080/01436597.2013.824657.

Mustafa, D., Altz-Stamm, A. & Scott, L., 2016. Water User Associations and the Politics of Water in Jordan. *World Development*, 79, C: 164–76.

Mustafa, D., Anwar, N. & Sawas, A., 2019. Gender, Global Terror, and Everyday Violence in Urban Pakistan. *Political Geography*, 69, no. 1: 54–64.

Mustafa, D. & Wrathall, D., 2011. Indus Basin Floods of 2010: Souring of a Faustian Bargain?. *Water Alternatives*, 4, no. 1: 72–85.

Mustafa, D., Gioli, G., Qazi, S., Rehman, A. & Zahoor, R., 2015. Gendering Flood Early Warning Systems: The Case of Pakistan. *Environmental Hazards*, 14, no. 4: 312–28.

Mustafa, D., Klassert, C., Yoon, J., Gawel, E., Sigel, K., Klauer, B., Talozi, S., Lachaut, T., Selby, P., Knox, S., Gorelick, S., Tilmant, A., Harou, J., Medellin-Azuara, J., Rajsekhar, D., Avisse, N. & Zhang, H., 2017. Spatial Analysis of Private Tanker Water Markets in Jordan: Using a Hydroeconomic Multi-agent Model to Simulate Non-observed Water Transfers. *EGU General Assembly Conference Abstracts*, p. 16483.

Naveed, A., 2019. Power Defaulters Owe Rs 244.8b in Outstanding Bills. *The Express Tribune*. https://tribune.com.pk/story/1963033/2-power-defaulters-owe-rs244-8b-outstanding-bills/.

Nesbit, J., 2018. *How Droughts and Die-offs, Heat Waves and Hurricanes are Converging on America: This is the Way the World Ends*. New York, NY: St Martin Press.

Ousat, A., 2015. Illegal Connections? Severe Water Crisis Hits DHA Clifton. *Express Tribune*. http://tribune.com.pk/story/835628/illegal-connections-severe-water-crisis-hits-dha-clifton/.

PILDAT, 2011. Interprovincial Water Issues in Pakistan. www.pildat.org/publications/publication / WaterR /Inter-ProvincialWaterIssuesinPakistan-BackgroundPaper.pdf.

Qutub, S. A., Saleemi, A. R., Reddy, M. S., Char, N. V. V., Gyawali, D., Sajjadur Rasheed, K. B. & Nickum, J. E., 2003. *Water Sharing Conflicts within Countries and Possible Solutions*. Honolulu, HI: Global Environment and Energy in the 21st Century. Retrieved on 6 July 2020 from URL: http://www.gee-21.org/publications/Water-Sharing-Conflicts-within-Countries-and-Possible-Solutions.pdf.

Rahman, P., 2008. *Water Supply in Karachi: Situation/Issues, Priority Issues and Solutions*. Karachi: Orangi Pilot Project-Training and Research Institute. http://labs.tribune.com.pk/aqua-final/perween -rahman-water-study.pdf.

Shah, T., 2007. The Groundwater Economy of South Asia: An Assessment of Size, Significance and Socio- Ecological Impacts. *The Agricultural Groundwater Revolution: Opportunities and Threats to Development*, 7–36.

S.E., 2015. Superintending Engineer. Punjab Irrigation Department, personal communication.

S.G, 2015. Government of Sindh. PC-II Proforma for a Feasibility Study for Supplying Ecological Flow in Manchar Lake, *Irrigation Department*.

Swyngedouw, E., 1999. Modernity and Hybridity: Nature, Regeneracionismo, and the Production of the Spanish Waterscape, 1890–1930. *Annals of the Association of American Geographers*, 89, no. 3: 443–65.

Swyngedouw, E., 2004. *Social Power and the Urbanization of Water*. Oxford: Oxford Geographical and Environmental Studies Series.

van Steenbergen, F., 1997. Understanding the Sociology of Spate Irrigation: Cases from Balochistan. *Journal of Arid Environments*, 35, no. 2: 349–65.

van Steenbergen, F., Kaisarani, A. B., Khan, N. U. & Gohar, M. S., 2015. A Case of Groundwater Depletion in Balochistan, Pakistan: Enter into the Void. *The Journal of Hydrology: Regional Studies*, 4, no. 1: 36–47.

Wescoat, J. L., Halvorson, S. & Mustafa, D., 2000. Water Management in the Indus Basin of Pakistan: A Half-Century Perspective. *International Journal of Water Resources Development*, 16, no. 3: 391–406.

White House, Department of Interior Panel on Waterlogging & Salinity in West Pakistan, 1964. *Report on Land and Water Development in the Indus Plain*. Washington DC: US Government Printing Press.

Wolf, A., 2002. *Conflict prevention and resolution in water systems*. Cheltenham: Edward Elgar Publishers.

Yu, W., Savitsky, A., Alford, D. L., Brown, C., Debowicz, D. J., Robinson, S., Wescoat Jr., J. L. & Yang, Y. E., 2013. *The Indus Basin of Pakistan: The Impacts of Climate Risks on Water and Agriculture*. Washington DC: World Bank.

Zaidi, A. S., 2014. Karachi as a Province. *Dawn*. Retrieved on 28 December 2015 from URL: http://www.dawn.com/news/1079781.

Zeitoun, M. & Warner, J., 2006. Hydro-hegemony – A Framework for Analysis of Transboundary Water Conflicts. *Water Policy*, 8, no. 5: 435–60.

Zeitoun, M. & Mirumachi, N., 2008. Transboundary Water Interaction I: Soft Power Underlying Conflict and Cooperation. *Int Environ Agreements*, 8: 297. https://doi.org/10.1007/s10784-008-9083-5.

Index

Abee Jarhiyyat 64
absolute water scarcity 18
accord 16, 23–5, 28–34
agricultural communities 51
Allan, S. 18
Anwar, N. H. 90
average flows 30–1, 33

Balochistan 13, 15–16, 18–19, 26–7, 31–2, 38–9, 43–4, 47–8, 55, 66, 82
   groundwater conflict in 47–50
barrages 30, 58–9, 67, 70
Bhagwandas 89, 93–4
Bhutto, Zulfiqar Ali 74
Bilal Colony 89–90
boats 65–6, 71, 76–7
bureaucracy 25–6, 52–3

Canal and Drainage Act 50
canals 14, 32, 38, 44–6, 50, 53, 58, 60, 66, 68–70, 74
Chashma-Jehlum link canal 24, 29
Chatterjee, P. 6
class 19, 57, 64, 81, 84, 88–9, 96–7, 101–2
   conflict 71, 90
Clifton Cantonment Board (CCB) 83
climate change 15
communities 18, 47–9, 53, 66, 72, 84, 93, 95–7, 101
   scale 86, 88, 97
conflict 1–3, 5–8, 10–11, 18–20, 22–3, 34, 40–1, 43–5, 48–51, 57–8, 65–7, 69–71, 73–5, 78–83, 87–90, 93–7, 99–103
   domestic water supply 81–98
contemporary Sindhi nationalists 28
Coriolis effect 60

Daily Nawa-e-Waqt 64
Dalin, C. 18
dams 11, 15, 21, 34–40, 59, 64–5, 69–70, 79

Dawn 65
decentralization 100
de facto privatization 82
domestic water supply 81–98
   physical and institutional context of 82–3
domestic water supply conflict 73
drainage 14, 19, 58, 60–1, 65, 70, 79, 100
   of flood waters 60–1
   as hazard 61–3
   management 63–7

electricity 33, 38, 49, 72, 88, 98
environment relations 7–8, 10, 41, 103
ethnicities 26
ethno-nationalist parties 27
everyday water conflict 11
expertise 3

famine 43
Farmer Association of Pakistan 17
farmers 1, 16, 46–9, 51–4, 66, 74–7
   conflict 74, 77
Federal Flood Commission (FFC) 64
Federally Administered Tribal Areas (FATA) 16
First Information Report (FIR) 77
fisher communities 57, 65–6, 74–6
fisherfolk 1, 18, 72, 74–6, 100
fisherfolk-farmer conflict 74
Fisher Folk Forum 17
fishermen 65, 74–5
floods 9–10, 18–19, 31, 57–8, 61, 63–5, 68–70, 77–80, 99–101
   conflicts 57, 67
   conflicts in Sindh and Punjab 67–71
   flows 30, 68
   fractal scalar politics in 63–7
   hazard 6, 18, 57, 63–6, 70, 77–8
   historico-physical geography of 58–61

management 57, 59, 65, 69–70, 78, 100, 102
    waters 60–1
Foucault 6
fractal concept 5

geographical scales 1, 4, 19, 103
Gilmartin, D. 6
glacial lake outburst floods (GLOF) 61
groundwater conflict 43
    in Balochistan 43, 47
groundwater development 43
Gujjar Nala 87, 93–6

hazardousness 9–11
hazards 3, 5–7, 9–11, 21–2, 38–40, 58, 61, 65, 78–81, 93, 100, 103
hazardscapes 8–11, 19–21, 24, 43, 77, 81, 93, 100–1
heads of water courses 45–6
Henderson, G. 10
Hewitt, K. 6
highlands of Balochistan 15, 43–4
Hoefle, S. W. 5
household scale water users 97
hydro-hazardscapes 1–20, 97, 98
hydropolitics 3, 100, 102–3

India 12, 16, 18, 23–4, 27–9, 33, 36–7, 41, 64–6, 81, 84, 89, 99–100
    water diversions 41
    water managers 64
Indus Basin 14–15, 33, 36, 40, 59–60, 68, 79, 99
    in Pakistan 14
Indus River 2, 11–13, 15–16, 23–4, 38–9, 58, 61–2
Indus River System 12, 14, 22–3, 40, 69
Indus Waters Treaty (IWT) 2, 29, 34, 65
international agencies 17
international-scale water policy 3
inter-provincial conflicts 18, 27, 41
Inter Provincial Water Accord 24
inter-provincial water allocations 16, 25, 34
inter-provincial water conflict 2, 21–41, 99–100
    agreements, allocations and royalties in 28–34
    historical antecedents of 23–5

Kalabagh or death 34–40
    mitigating 40–1
    politico-historic context of 25–8
inter-provincial water disputes 23, 25
inter-provincial water management 24
inter-provincial water politics 22, 25, 28
interprovincial water thefts 34
intra-provincial water distribution 22
inundation zones 59–60, 69, 79
irrigation department 50, 52, 68, 70, 75
irrigation water 6–7, 53, 102
    conflicts 53–5

Jonas, A. E. G. 5

Kalabagh Dam (KBD) 17–18, 21, 30, 34–41, 99
Karachi 16, 23, 81–99, 101–2
Karachi Bulk Water Supply Scheme 83
Karachi's water problems 83, 96
Karachi Water and Sewerage Board (KWSB) 16, 82–4, 86, 95, 102
Karachi Water and Sewerage Board Act 1996 82
karezes 44, 47–9, 55, 101
    system 44, 46–9, 54, 101
    water 48–9
Karner, Milan 51
Karpouzoglou, T. 9
Khan Mohammad Mallah village 66, 71–2, 75
Khyber Pakhtunkhwa (KP) 13
Kirthar Canal 31–2
Kotri Barrage 31

land 14–15, 44, 46, 48, 66, 68, 70, 73, 75–6, 78, 96, 102
landscape 9–10
large farmers 18, 50, 52–4, 76, 101
large-scale water storage 16
Latour, B. 9
Leitner, H. 5
link canals 23–4, 29
Linton, J. 1
local hazardscapes, hazards 57–80
local irrigation water conflicts 53
local-level water conflict 43, 44
local-scale conflicts 19, 43, 46, 66, 77, 99, 101
local-scale customary water rights 50

local scale of water management 52
local-scale politics 18, 97
local-scale surface water 101
local-scale water conflict
    canal and karez irrigation, historico-institutional context 44–6
    surface and groundwater, Pakistan 43–55
    surface irrigation water conflict 50–3
local-scale water management system 101

Manchar Lake 61, 63, 65–6, 71, 74
Marston, S. 4–5
Michel, A. A. 23
Miller, B. 5
Mirumachi, N. 2
Moga 46
Mohenjo Daro 11
Moore, S. M. 3–4, 7, 9, 100
Mustafa, D. 9–10, 50–2, 59, 90
Mutahida Qaumi Movement (MQM) 27, 84, 89–90, 96
Muzzaffargarh Canal 69

nationalist hazardscapes, inter-provincial water conflict 21–41
national-level water scarcity 36
national scale of flood management 65
national scale of Pakistan 40, 66
national water scarcity risk 40

oil exploration 75–6
Orangi 84, 86

Pakistan 1–2, 12–28, 30, 34–8, 40–1, 43–4, 50–5, 57–8, 60, 62–6, 68–70, 81, 90, 92, 99–100, *see also individual entries*
    physical geography of water, Indus Basin 12–15
    water governance context 15–18
    water scarce/vulnerable country 12–18
Pakistan Commission for Indus Water 65
Pakistan Council for Research in Water Resources (PCRWR) 21
Pakistani Punjab 65

Pakistan Meteorological Department (PMD) 78
Pakistan Muslim League (PML) 27
Pakistan People's Party 27–8
Pakistan Tehrik-e-Insaf (PTI) 27
Pakistan Water Partnership 17
parcel of land 44, 75
piped water system 84, 102
political conflicts 50–1, 96
pollution hazard
    downstream conflicts from 71–7
    farmers *vs.* fishermen 74–7
power 6–8, 88–96
    relations 7, 9, 11, 19, 100–2
privatization 83–8
provincial scale 3, 24, 38, 40, 50, 73, 100
    water bureaucracies 53–4
pukki warabandi 50
Punjab 13, 16–19, 22–33, 36–40, 43–5, 50, 55, 58–9, 61, 64–7, 69–70, 99–100
    flood conflicts in 67–71
Punjab Irrigation Department 30, 67–9
Punjabi water bureaucracy 39

Qutub, S. A. 5

Rahman, P. 84
Ravi River 60
real water scarcity 33
religion-based water conflict 102
reverse osmosis (RO) plants 71, 72
Right Bank Outfall Drain (RBOD) project 62, 63, 76–8
rivers 1, 11, 13–14, 22–4, 29, 39, 57–60, 64, 70, 77

Saeed, Hafiz 64
safe design capacity 59, 67, 79
saline groundwater 14, 37, 62
saline water 93–4
Salinity Control and Reclamation Projects (SCARP) 62
Sidhnai Barrage 60, 68
signature water conflicts 2
Sindh 13–14, 16, 18, 22–32, 34, 36–41, 44, 55, 57–8, 60–2, 65–7, 70–1, 99–100

conflict over drainage and water
  quality 65–7
 flood conflicts in 67–71
Sindh Fisheries Act 1980, Clause 7 75
Sindh Fisheries (Amendment) Act
  2011 74
smaller farmers 53
social capital 55
social consciousness 8
social power 6
spatial scales 5–6, 10–11, 18
sub-national inter-provincial-scale water
  conflict 43
sub-national scale hydropolitics 3–4
sub-national scale water conflicts 3,
  40–1, 55
suction pumps 87–8, 102
surface irrigation water conflict 50–3
surplus water 29, 51
sweet water 75
Swyngedouw, E. 7

tankers 84, 86–8, 93–4, 96, 102
Tarbela Dam 23, 33, 35
Taunsa Barrage 69
technocracy 3
transboundary water conflict 1–2, 5
Transboundary Waters Interaction Nexus
  (TWINS) 3
Trimmu-Panjnad Link Canal 29
trust 38–9, 80, 99
tubewells 14, 43–4, 46, 48–9, 62, 101

UN Economic and Social Commission for
  West Asia (UNESCWA) 7
Urban Pakistan 101–2
urban water conflict 98
US Mexico Treaty, 1944 1

valves 87–91, 95–6
Vij, S. 9
village water courses 19, 46, 50, 99

Warner, J. 2
water 1–12, 15, 17–19, 21–4, 28–33, 36,
  38–41, 44–7, 49–55, 57–8, 60, 69,
  71–2, 75–6, 78, 81–103, *see also*
  individual entries

access 5, 51, 97
courses 46, 53
delivery 29, 88, 90
development 7
dispute 51
distribution conflicts 18
distribution system 44, 46
flows 33, 64, 71, 86, 100
governance context 15–18
infrastructure 21, 47
management 16, 24, 34, 40, 50, 52–3,
  65, 103
politics 18, 20
pollution 19, 57–8, 77–8
resources 15, 17, 21, 33
rights 19, 44
scarcity 18–19, 21–2, 41, 48, 51, 93,
  99, 101–2
stoppage 99–100
supply 12, 32, 52, 55, 82–4, 88, 92
systems 12
tankers 83–4, 93, 102
theft 52
vendors 86, 94
Water Accord 28–9
water allocations 1, 23–4, 28, 30
Water and Power Development Authority
  (WAPDA) 16, 79
Water and Sanitation Agencies
  (WASA) 16
water conflict 1–6, 8–12, 15, 18–20,
  41, 43–4, 49–50, 54–5, 73, 97,
  99–103
 actuates 81
 in hazardous world 8–12
 literature 43
 management 41
waterlogging 62, 100
water-related conflicts 48
water-related hazards 9, 78
waterscape 7–9
water user associations (WUAs) 52–4,
  101
Water Wars thesis 2
Wescoat, J. L. 63
Wrathall, D. 59

Zeitoun, M. 2